BIG MIND

GEOFF MULGAN

BIG MIND

HOW COLLECTIVE INTELLIGENCE CAN CHANGE OUR WORLD

PRINCETON UNIVERSITY PRESS

PRINCETON AND OXFORD

Published by Princeton University Press
41 William Street, Princeton, New Jersey 08540

In the United Kingdom: Princeton University Press
6 Oxford Street, Woodstock, Oxfordshire OX20 1TR

press.princeton.edu

Jacket design by Karl Spurzem
Vectors courtesy of Shutterstock

ISBN 978-0-691-17079-4
Library of Congress Control Number: 2017954117

British Library Cataloging-in-Publication Data is available

This book has been composed in Adobe Garamond Pro

Printed on acid-free paper. ∞

Printed in the United States of America

1 3 5 7 9 10 8 6 4 2

CONTENTS

Part IV

Collective Intelligence as Expanded Possibility   215

PREFACE

THIS BOOK HAS BEEN SEVERAL DECADES in the making. It grows out of both experience and research. The experience has been the practical work of trying to help businesses, governments, and nongovernmental organizations (NGOs) solve problems, use technologies, and act smarter. Alongside that, much of my research and writing has essentially been about how thought happens on a large scale. *Communication and Control: Networks and the New Economies of Communication* (Blackwell, 1990) was about the nature of the new networks being made possible by digital technologies and the kinds of control they brought with them. It showed how networks could both empower and disempower (and was intended as a corrective to the hopes that networks would automatically usher in an era of greater democracy, equality, and freedom). *Connexity: How to Live in a Connected World* (Harvard Business Press, 1997) was a more philosophical essay about the morality of a connected world, and the types of people and character that would be needed in a networked environment. *Good and Bad Power: The Ideals and Betrayals of Government* (Penguin, 2005) and *The Art of Public Strategy: Mobilizing Power and Knowledge for the Common Good* (Oxford University Press, 2009) were about how the state could use its unique powers to the greatest good, including mobilizing and working with the brainpower of citizens. *The Locust and the Bee: Predators and Creators in Capitalism's Future* (Princeton University Press, 2013) set out a new agenda for economics, suggesting how economies could expand collective intelligence and creative potential while reining in predatory tendencies.

What follows here builds on each of these, weaving them into what I hope is both a convincing theory and useful guide.

The ideas draw on my previous work, but have also benefited greatly from many conversations, readings, and arguments. I owe a significant debt to my colleagues at Nesta, particularly Stefana Broadbent, Tom Saunders, John Loder, Francesca Bria, Stian Westlake, and Zosia Poulter (for the diagrams). My colleagues at Harvard's Ash Center were generous in terms of their time and engagement, especially Mark Moore, who provided extensive and useful comments, and Jorrit de Jong. I am particularly grateful to the Ash Center for having given me the chance to be a senior visiting scholar over three years, from 2015 to 2018, to try out some of the ideas explored in this book. Also at Harvard, Roberto Mangabeira Unger and Howard Gardner once again offered invaluable stimulus.

In addition, I want to extend my appreciation for the intellectual collaborators in and around GovLab at New York University, especially Beth Noveck and Stefan Verhuist. Marta Struminska from Warsaw provided useful early comments, as did Rushanara Ali, Lynne Parsons, Robin Murray, Soh Yeong Roh, Gavin Starks, Sarah Savant, Vaughn Tan, and Francois Taddei. Colin Blakemore and Mattia Gallotti organized a fascinating conference at Nesta on collective intelligence in 2015. Mattia went on to offer helpful and detailed comments on a draft, for which I'm hugely grateful. I've also benefited greatly from support and insights from others in and around this field, including Karim Lakhani at Harvard, Tom Malone at the Massachusetts Institute of Technology, and Joshua Ramo. I owe thanks as well to Julia Hobsbawm for giving me the opportunity to test out some of the arguments on a group that included the historian Simon Schama and journalist David Aaronovich, to Luciano Floridi for letting me air some of the ideas in the journal *Philosophy and Technology*, and Jens Wandel and Gina Lucarelli at the UN Development Program for the chance to put them into practice in the field of development.

**BIG
MIND**

Introduction

Collective Intelligence as a Grand Challenge

THERE ARE LIBRARIES FULL OF BOOKS on individual intelligence, investigating where it comes from, how it manifests, and whether it's one thing or many. Over many years, I've been interested in a less studied field. Working in governments and charities, businesses and movements, I've been fascinated by the question of why some organizations seem so much smarter than others—better able to navigate the uncertain currents of the world around them. Even more fascinating are the examples of organizations full of clever people and expensive technology that nevertheless act in stupid and self-destructive ways.

I looked around for the theories and studies that would make sense of this, but found little available.[1] And so I observed, assessed, and drew up hypotheses.

I was helped in this study by having been trained in things digital, completing a PhD in telecommunications. Digital technologies can sometimes dumb people down. But they have the virtue of making thought processes visible. Someone has to program how software will process information, sensors will gather data, or memories will be stored. All of us living in a more pervasively digital age, and those of us who have to think digitally for our work, are inevitably more sensitive to how intelligence is organized, where perhaps in another era we might have thought it a fact of nature, magical, and mysterious.

The field that led me to has sometimes been given the label *collective intelligence*. In its narrow variants, it's mainly concerned with how groups of people collaborate together online. In its broader variants it's about how all kinds of intelligence happen on large scales. At its extreme, it encompasses the whole of human civilization and culture, which constitutes the collective intelligence of our species, passed down imperfectly through books and schools, lectures and demonstrations, or by parents showing children how to sit still, eat, or get dressed in the morning.

My interest is less ambitious than this. I'm concerned with the space between the individual and the totality of civilization—an equivalent to the space in biology between individual organisms and the whole biosphere. Just as it makes sense to study particular ecologies—lakes, deserts, and forests—so it also makes sense to study the systems of intelligence that operate at this middle level, in individual organizations, sectors, or fields.

Within this space, my primary interest is narrower still: How do societies, governments, or governing systems solve complex problems, or to put it another way, how do collective problems find collective solutions?

Individual neurons only become useful when they're connected to billions of other neurons. In a similar way, the linking up of people and machines makes possible dramatic jumps in collective intelligence. When this happens, the whole can be much more than the sum of its parts.

Our challenge is to understand how to do this well; how to avoid drowning in a sea of data or being deafened by the noise of too much irrelevant information; how to use technologies to amplify our minds rather than constrain them in predictable ruts.

What follows in this book is a combination of description and theory that aims to guide design and action. Its central claim is that every individual, organization, or group could thrive more successfully if it tapped into a bigger mind—drawing on the brainpower of other people and machines. There are already some three billion people connected online and over five billion connected machines.[2] But making the most of them requires careful attention to methods, avoidance of traps, and investment of scarce resources.[3] As is the case with the links between neurons in our brain, successful thought depends on structure and organization, not just the number of connections or signals.

This may be more obvious in the near future. Children growing up in the twenty-first century take it for granted that they are surrounded by sensors and social media, and their participation in overlapping group minds—hives, crowds, and clubs—makes the idea that intelligence resides primarily in the space inside the human skull into an odd anachronism. Some feel comfortable living far more open and transparent lives than their parents, much more part of the crowd than apart.

The great risk in their lifetimes, though, is that collective intelligence won't keep up with artificial intelligence. As a result, they may live in a future where extraordinarily smart artificial intelligence sits amid often-inept systems for making the decisions that matter most.

To avoid that fate we need clear thinking. For example, it was once assumed that crowds were by their nature dangerous, deluded, and cruel. More recently the pendulum swung to an opposite assumption: that crowds tend to be wise. The truth is subtler. There are now innumerable examples that show the gains from mobilizing more people to take part in observation, analysis, and problem solving. But crowds, whether online or off-line, can also be foolish and biased, or overconfident echo chambers. Within any group, diverging and conflicting interests make any kind of collective intelligence both a tool for cooperation and a site for competition, deception, and manipulation.

Taking advantage of the possibilities of a bigger mind can also bring stark vulnerabilities for us as individuals. We may, and often will, find our skills and knowledge quickly superseded by intelligent machines. If our data and lives become visible, we can more easily be exploited by powerful predators.

For institutions, the rising importance of conscious collective intelligence is no less challenging, and demands a different view of boundaries and roles. Every organization needs to become more aware of how it observes, analyses, remembers, and creates, and then how it learns from action: correcting errors, sometimes creating new categories when the old ones don't work, and sometimes developing entirely new ways of thinking. Every organization has to find the right position between the silence and the noise: the silence of the old hierarchies in which no one dared to challenge or warn, and the noisy cacophony of a world of networks flooded by an infinity of voices. That space in between becomes meaningful only when organizations learn how to select and cluster with the right levels of granularity—simple enough but not simplistic; clear but not crude; focused but not to the extent of myopia. Few of our dominant institutions are adept at thinking in these ways. Businesses have the biggest incentives to act more intelligently, and invest heavily in hardware and software of all kinds. But whole sectors repeatedly make big mistakes, misread their environments, and harvest only a fraction of the know-how that's available in their employees and customers. Many can be extremely smart within narrow parameters, but far less so when it comes to the bigger picture. Again and again, we find that big data without a big mind (and sometimes a big heart) can amplify errors of diagnosis and prescription.

Democratic institutions, where we, together, make some of our most important decisions, have proven even less capable of learning how to learn. Instead, most are frozen in forms and structures that made sense a century or two ago, but are now anachronisms. A few parliaments and cities are trying to harness the collective intelligence of their citizens. But many democratic institutions—parliaments, congresses, and parties—look dumber than the societies they serve. All too often the enemies of collective intelligence are able to capture public discourse, spread misinformation, and fill debates with distractions rather than facts.

So how can people think together in groups? How might they think and act more successfully? How might the flood of new technologies available to help with thinking—technologies for watching, counting, matching, and predicting—help us together solve our most compelling problems?

In this book, I describe the emerging theory and practice that points to different ways of seeing the world and acting in it. Drawing on insights from many disciplines, I share concepts with which we can make sense of how groups think, ideas that may help to predict why some thrive and others falter, and pointers as to how a firm, social movement, or government might think more successfully, combining the best of technologies with the best of the gray matter at its disposal.

I sketch out what in time could become a full-fledged discipline of collective intelligence, providing insights into how economies work, how democracies can be reformed, or the difference between exhilarating and depressing meetings. Hannah Arendt once commented that a stray dog has a better chance of surviving if it's given a name, and in a similar way this field may better thrive if we use the name collective intelligence to bring together many diverse ideas and practices.

The field needs to be both open and empirical. Just as cognitive science has drawn on many sources—from linguistics to neuroscience, psychology to anthropology—to understand how people think, so will a new discipline concerned with thought on larger scales need to draw on many disciplines, from social psychology to computer science, economics to sociology, and use these to guide practical experiments. Then, as the new discipline emerges—and is hopefully helped by neighboring disciplines rather than attacked for challenging their boundaries—it will need to be closely tied into practice: supporting, guiding, and learning from a community of practitioners working to design as well as operate tools that help systems think and act more successfully.

Collective intelligence isn't inherently new, and throughout the book I draw on the insights and successes of the past, from the nineteenth-century designers of the *Oxford English Dictionary* (*OED*) to the Cybersyn project in Chile, from Isaac Newton's *Principia Mathematica* to the National Aeronautics and Space Administration (NASA), from Taiwanese democracy to Finnish universities, and from Kenyan web platforms to the dynamics of football teams.

In our own brains, the ability to link observation, analysis, creativity, memory, judgment, and wisdom makes the whole much more than the sum of its parts. In a similar way, I argue that assemblies that bring together many elements will be vital if the world is to navigate some of its biggest challenges, from health and climate change to migration. Their role will be to orchestrate knowledge and also apply much more systematic methods to knowledge about that knowledge—including metadata, verification tools, and tags, and careful attention to how knowledge is used in practice. Such assemblies are multiplicative rather than additive: their value comes from how the elements are connected together. Unfortunately they remain rare and often fragile.

To get at the right answers, we'll have to reject appealing conventional wisdoms. One is the idea that a more networked world *automatically* becomes more intelligent through processes of organic self-organization. Although this view contains important grains of truth, it has been deeply misleading.[4] Just as the apparently free Internet rests on energy-hungry server farms, so does collective intelligence depend on the commitment of scarce resources. Collective intelligence can be light, emergent, and serendipitous. But it more often has to be consciously orchestrated, supported by specialist institutions and roles, and helped by common standards. In many fields no one sees it as their role to make this happen, as a result of which the world acts far less intelligently than it could.

The biggest potential rewards lie at a global level. We have truly global Internet and social media. But we are a long way short of a truly global collective intelligence suitable for solving global problems—from pandemics to climate threats, violence to poverty. There's no shortage of interesting pilots and projects. Yet we sorely lack more concerted support and action to assemble new combinations of tools that can help the world think and act at a pace as well as scale commensurate with the problems we face. Instead, in far too many fields the most important data and knowledge are flawed and fragmented, lacking the organization that's needed to make them easy to access and use, and no one has the means or capacity to bring them together.

Perhaps the biggest problem is that highly competitive fields—the military, finance, and to a lesser extent marketing or electoral politics—account for the majority of investment in tools for large-scale intelligence. Their influence has shaped the technologies themselves. Spotting small variances is critical if your main concern is defense or to find comparative advantage in financial markets. So technologies have advanced much further to see, sense, map, and match than to understand. The linear processing logic of the Turing machine is much better at manipulating inputs than it is at creating strong models that can use the inputs and create meanings. In other words, digital technologies have developed to be good at answers and bad at questions, good at serial logic and poor at parallel logic, and good at large-scale processing and bad at spotting nonobvious patterns.

Fields that are less competitive but potentially offer much greater gains to society—such as physical and mental health, environment, and community—have tended to miss out, and have had much less influence on the direction of technological change.[5] The net result is a massive misallocation of brainpower, summed up in the lament of Jeff Hammerbacher, the former head of data at Facebook, that "the best minds of my generation are thinking about how to make people click ads."

The stakes could not be higher. Progressing collective intelligence is in many ways humanity's grandest challenge since there's little prospect of solving the other grand challenges of climate, health, prosperity, or war without progress in how we think and act together.

We cannot easily imagine the mind of the future. The past offers clues, though. Evolutionary biology shows that the major transitions in life— from chromosomes to multicellular organisms, prokaryotic to eukaryotic cells, plants to animals, and simple to sexual reproduction—all had a common pattern. Each transition led to a new form of cooperation and interdependence so that organisms that before the transition could replicate independently, afterward could only replicate as "part of a larger whole."

Each shift also brought with it new ways of both storing and transmitting information.

It now seems inevitable that our lives will be more interwoven with intelligent machinery that will shape, challenge, supplant, and amplify us, frequently at the same time. The question we should be asking is not *whether* this will happen but rather *how* we can shape these tools so that they shape us well—enhancing us in every sense of the word and making us more of what we most admire in ourselves. We may not be able to avoid a world of virtual reality pornography, ultrasmart missiles, and spies. But we can create a better version of collective intelligence alongside these—a world where, in tandem with machines, we become wiser, more aware, and better able to thrive and survive.

The Structure of the Book

The rest of this book is divided into four main sections.

The first section (chapters 1 and 2) maps out the issue and explains what collective intelligence is. I offer illustrations of collective intelligence in practice, outline ways of thinking about it, and describe some of the most interesting contemporary examples.

The next section focuses on how to make sense of collective intelligence (chapters 3 to 10). It provides a theoretical framework that describes the functional elements of intelligence and how they are brought together, how collectives are formed, and how intelligence struggles with its enemies.

Chapters 11 to 17 then look at collective intelligence in the wild along with the implications of the theories for specific fields: the organization of meetings and places, business and the economy, democracy, the university, social change, and the new digital commons. In each case, I show how thinking about collective intelligence can unlock new perspectives and solutions. Finally, in chapter 18, I pull the themes together and address the politics of collective intelligence, demonstrating what progress toward greater collective wisdom might look like.

PART I

What Is Collective Intelligence?

IN THIS FIRST SECTION, I explain what collective intelligence means in practice and how we can recognize it in the world around us, helping us to plan a journey, diagnose an illness, or track down an old friend.

It's an odd paradox that ever more intelligent machines can be found at work within systems that behave foolishly. Despite the unevenness of results, however, there are many promising initiatives to support intelligence on a large scale that have drawn on a cascade of advances in computing, from web science to machine learning. These range from household names like Google Maps and Wikipedia to more obscure experiments in math and chess. Connecting large numbers of machines and people makes it possible for them to think in radically new ways—solving complex problems, spotting issues faster, and combining resources in new ways.

How to do this well is rarely straightforward, and crowds aren't automatically wise. But we are beginning to see subtler forms of what I call *assemblies* emerge. These bring together many elements of collective intelligence into a single system. They show how the world could think on a truly global scale, tracking such things as outbreaks of disease or the state of the world's environments, and feeding back into action. For example, an observatory that spots global outbreaks of Zika can predict how the virus might spread and guide public health services to direct their resources to contain any outbreaks. Within cities, combining large data sets can make it easier to spot which buildings are at most risk of fires or which hospital patients are most at risk of becoming sick, so that government can be more adept at predicting and preventing rather than curing and fixing.

These ways of organizing thought on a large scale are still in their infancy. They lack a convincing guiding theory and professional experts who know the tricks of the trade. In many cases, they lack a reliable economic base. Yet they suggest how in the future, almost every field of human activity could become better at harnessing information and learning fast.

- 1 -

The Paradox of a Smart World

WE LIVE SURROUNDED BY NEW WAYS of thinking, understanding, and measuring that simultaneously point to a new step in human evolution and an evolution beyond humans.

Some of the new ways of thinking involve data—mapping, matching, and searching for patterns far beyond the capacity of the human eye or ear. Some involve analysis—supercomputers able to model the weather, play chess, or diagnose diseases (for example, using the technologies of firms like Google's DeepMind or IBM's Watson). Some pull us ever further into what the novelist William Gibson described as the "consensual hallucination" of cyberspace.[1]

These all show promise. But there is a striking imbalance between the smartness of the tools we have around us and the more limited smartness of the results. The Internet, World Wide Web, and Internet of things are major steps forward in the orchestration of information and knowledge.

Yet it doesn't often feel as if the world is all that clever. Technologies can dumb down as well as smarten up.[2] Many institutions and systems act much more stupidly than the people within them, including many that have access to the most sophisticated technologies. Martin Luther King Jr. spoke of "guided missiles but misguided men," and institutions packed with individual intelligence can often display collective stupidity or the distorted worldview of "idiots savants" in machine form. New technologies bring with them new catastrophes partly because they so frequently outstrip our wisdom (no one has found a way to create code without also creating bugs, and as the French philosopher Paul Virilio put it, the aircraft inevitably produces the air disaster).

In the 1980s, the economist Robert Solow commented, "You can see the computer age everywhere but in the productivity statistics." Today we might say again that data and intelligence are everywhere—except in the productivity statistics, and in many of the things that matter most. The financial crash of the late 2000s was a particularly striking example. Financial institutions that had spent vast sums on information technologies failed to understand what was happening to them, or understood the data but not what lay behind the data, and so brought the world to the brink of economic disaster.[3] In the 1960s and 1970s, the Soviet government had at its disposal brilliant minds and computers, but couldn't think its way out of stagnation. During the same period, the US military had more computing power at its disposition than any other organization in history, but failed to understand the true dynamics of the war it was fighting in Vietnam. A generation later the same happened in Iraq, when a war was fought based on a profound error of intelligence launched by the US and UK governments with more invested than any other countries in the most advanced intelligence tools imaginable. Many other examples confirm that having smart tools does not automatically lead to more intelligent results.

Health is perhaps the most striking example of the paradoxical combination of smart elements and often-stupid results. We now benefit from vastly more access to information on diseases, diagnoses, and treatments on the Internet. There are global databases of which treatments work; detailed guidance for doctors on symptoms, diagnoses, and prescriptions; and colossal funds devoted to pushing the frontiers of cancer, surgery, or pharmaceuticals.

But this is far from a golden age of healthy activity or intelligence about health. The information available through networks is frequently misleading (according to some research, more so than face-to-face advice).[4] There are well over 150,000 health apps, yet only a tiny fraction can point to any evidence that they improve their users' health. The dominant media propagate half-truths and sometimes even lies as well as useful truths. And millions of people make choices every day that clearly threaten their own health. The world's health systems are in many ways pioneers of collective intelligence, as I will show later, but much doesn't work well. It's estimated that some 30 to 50 percent of antibiotic prescriptions are unnecessary, 25 percent of medicines in circulation are counterfeit, somewhere between 10 and 20 percent of diagnoses are incorrect, and each year 250,000 die in the United States alone because of medical error (the third leading cause of death there).[5]

In short, the world has made great strides in improving health and has accumulated an extraordinary amount of knowledge about it, yet still has a long way to go in orchestrating that knowledge to best effect.

Similar patterns can be found in many fields, from politics and business to personal life: unprecedented access to data, information, and opinions, but less obvious progress in using this information to guide better decisions. We benefit from a cornucopia of goods unimaginable to past generations, yet still too often spend money we haven't earned to buy things we don't need to impress people we don't like.

We have extraordinary intelligence in pockets, for specific, defined tasks. Yet there has been glacial, if any, progress in handling more complex, interconnected problems, and paradoxically the excitement surrounding new capacities to sense, process, or analyze may distract attention from the more fundamental challenges.[6]

In later chapters, I address what true collective intelligence would look like in some of the most important fields. How could democracy be organized differently if it wanted to make the most of the ideas, expertise, and needs of citizens? Various experiments around the world suggest what the answers might be, but they baffle most of the professionals brought up in traditional politics. How could universities become better at creating, orchestrating, and sharing knowledge of all kinds? There are seeds of different approaches to be found, but also extraordinary inertia in the traditional models of three-year degrees, faculty hierarchies, lecture halls, and course notes. Or again, how could a city administration, or national government, think more successfully about solving problems like traffic congestion, housing shortages, or crime, amplifying the capabilities of its people rather than dumbing them down?

We can sketch plausible and achievable options that would greatly improve these institutions. In every case, however, the current reality falls far short of what's possible, and sometimes tools that could amplify intelligence turn out to have the opposite effect. Marcel Proust wrote that "nine tenths of the ills from which intelligent people suffer spring from their intellect." The same may be true of collective intelligence.[7]

- 2 -

The Nature of Collective Intelligence
in Theory and Practice

THE WORD *INTELLIGENCE* HAS A COMPLEX HISTORY. In medieval times, the intellect was understood as an aspect of our souls, with each individual intellect linked into the divine intellect of the cosmos and God.[1] Since then, understandings of intelligence have reflected the dominant technologies of the era. René Descartes used hydraulics as a metaphor for the brain and believed that animating fluids connected the brain to limbs. Sigmund Freud in the age of steam power saw the mind in terms of pressure and release. The age of radio and electrics gave us the metaphors of "crossed wires" and being "on the same wavelength," while in the age of computers the metaphors turned to processing and algorithmic thinking, and the brain as computer.[2]

There are many definitions of intelligence. But the roots of the word point in a direction that is rather different from these metaphors. Intelligence derives from the Latin word *inter*, meaning "between," combined with the word *legere*, meaning "choose." This makes intelligence not just a matter of extraordinary memory or processing speeds. Instead it refers to our ability to use our brains to know which path to take, who to trust, and what to do or not do. It comes close in this sense to what we mean by freedom.[3]

The phrase collective intelligence links this with a related idea. The word collective derives from *colligere*. This joins *col*, "together," and once again, *legere*, "choose." The collective is who we choose to be with, who we trust to share our lives with. So collective intelligence is in two senses a concept about choice: *who* we choose to be with and *how* we choose to act.

The phrase has been used in recent years primarily to refer to groups that combine together online. But it should more logically be used to describe any kind of large-scale intelligence that involves collectives choosing to be, think, and act together. That makes it an ethical as well as technical term, which also ties into our sense of conscience—a term that is now usually understood as individual, but is rooted in the combination of *con* (with) and *scire* (to know).

Possibility

We choose in a landscape of possibilities and probabilities. In every aspect of our lives we look out into a future of possible events, which we can guess or estimate, though never know for certain. Many of the tools I describe through the course of this book help us make sense of what lies ahead, predicting, adapting, and responding. We observe, analyze, model, remember, and try to learn. Although mistakes are unavoidable, repeated mistakes are unnecessary. But we also learn that in every situation, there are possibilities far beyond what data or knowledge can tell us—possibilities that thanks to imaginative intelligence, we can sometimes glimpse.

Groups

One of the first historical accounts of collective intelligence is Thucydides's description of how an army went about planning the assault on a besieged town. "They first made ladders equal in length to the height of the enemy's wall, which they calculated by the help of the layers of bricks on the side facing the town, at a place where the wall had accidentally not been plastered. A great many counted at once, and, although some might make mistakes, the calculation would be more often right than wrong; for they repeated the process again and again, and, the distance not being great, they could see the wall distinctly enough for their purpose. In this manner they ascertained the proper length of the ladders, taking as a measure the thickness of the bricks."[4]

Understanding how we work together—the collective part of collective intelligence—has been a central concern of social science for several centuries. Some mechanisms allow individual choices to be aggregated in a socially useful way without requiring any conscious collaboration or shared identity. This is the logic of the invisible hand of the market and some of the recent experiments with digital collective intelligence like Wikipedia. In other cases (such as communes, friends on vacation, or work teams), there is the conscious mutual coordination of people with relatively equal power, which usually involves a lot of conversation and negotiation. Loosely networked organizations such as Alcoholics Anonymous are similar in nature. In others (for instance, big corporations like Google or Samsung, ancient Greek armies, or modern global NGOs), hierarchy organizes cooperation, with a division of labor between different tiers of decision making.

Each of these produces particular kinds of collective intelligence. Each feels radically different, and works well for some tasks and not others. In some cases there is a central blueprint, command center, or plan—someone who can see how the pieces fit together and may end up as a new building, a business plan, or initiative. In other cases the intelligence is wholly distributed and no one can see the big picture in advance. But in most cases the individual doesn't need to know much about the system they're part of: they can be competent without comprehension.[5]

The detailed study of how groups work shows that we're bound together not just by interests and habit but also by meanings and stories. But the very properties that help a group cohere can also impede intelligence. These include shared assumptions that don't hold true, a shared willingness to ignore uncomfortable facts, groupthink, group feel, and mutual affirmation rather than criticism. Shared thought includes not only knowledge but also delusions, illusions, fantasies, the hunger for confirmation of what we already believe, and the distorting pull of power that bends facts and frames to serve itself. The Central Intelligence Agency informing President George H. W. Bush that the Berlin Wall wouldn't fall, just as the news was showing it doing just that; investment banks in the late 2000s piling into subprime mortgages when all the indicators showed that they were worthless; Joseph Stalin and his team ignoring the nearly ninety separate, credible intelligence warnings that Germany was about to invade in 1941—all are examples of how easily organizations can be trapped by their frames of thinking.

We succumb all too readily to illusions of control and optimism bias, and when in a crowd can suspend our sense of moral responsibility or choose riskier options we would never go for alone. And we like to have our judgments confirmed, behaving all too often like the Texas sharpshooter who sprays the walls with bullets and then draws the target around where they hit. These are just a few reasons why collective intelligence is so frequently more like collective stupidity.

They show why most groups face a trade-off between how collective they are and how intelligently they can behave. The more they bond, the less they see the world as it really is. Yet the most successful organizations and teams learn how to combine the two—with sufficient suspension of ego and sufficient trust to combine rigorous honesty with mutual commitment.

GENERAL AND SPECIFIC

How we think can then be imagined as running in a continuum from general, abstract intelligence to intelligence that is relevant to specific places, people, and times. At one extreme there are the general laws of physics or the somewhat less general laws of biology. There are abstracted data, standardized algorithms, and mass-produced products. Much of modernity has been built on an explosion of this kind of context-free intelligence. At the other end of the spectrum there is rooted intelligence—intelligence that understands the nuances of particular people, cultures, histories, or meanings, and loses salience when it's removed from them.

The first kinds of intelligence—abstract, standardized, and even universal—are well suited to computers, global markets, and forms of collective intelligence that are more about aggregation than integration. By contrast, the ones at the other extreme—like knowing how to change someone's life or regenerate a town—entangle multiple dimensions, and require much more conscious iteration and integration along with sensitivity to context.

COLLECTIVE INTELLIGENCE AND CONFLICT

The simplest way to judge individual intelligence is by how well it achieves goals and generates new ones.[6] But this is bound to be more complex for any large group, which is likely to have many different goals and often-conflicting interests.

This is obviously true in the economy, since information is usually hoarded and traded rather than shared. Societies try to design arrangements (including patents and copyrights) that reward people for both creating and sharing useful information, though, as I will show in chapter 17, the rise of economies based on information and knowledge has shifted the balance between private ownership and the commons, and led to an evident underproduction of informational commons. Even if this problem were fixed, though, there would still be unavoidable tensions thanks to conflicting interests.

Many collective actions that appear stupid may be intelligent for some of the people involved, such as a country setting out on a war it is unlikely to win (to shore up support for an insecure dictator), a bank taking apparently mad risks (which offer huge personal gains to a few at the top), or a religious community holding on to beliefs in the face of overwhelming contrary evidence (as the price for holding the community together).

Many traditions of social science have grappled with these issues: How can the principal (for instance, the public) ensure that their agent (say, government) really does act in their interests?

Seeing these issues through the lens of collective intelligence opens up the possibility that shared observation, reasoning, memory, and judgment will all increase the pressure to find mutually advantageous solutions. Think, for example, of a country coming out of civil war and strife. The ones that have done well intensify what we will see later as the hallmarks of successful collective intelligence: bringing facts and feelings to the surface in ways that are detached from interests; jointly deliberating about what is to be done and opening up alternative scenarios; discussing openly who is to be punished and who deserves restitution; and addressing memory openly through truth and reconciliation commissions.

In more everyday circumstances, one of the many roles of public institutions—from parliaments to businesses—is to turn fragmented, conflicting groups into something closer to a collective intelligence, able to find mutually satisfactory and legitimate answers through shared assessment, dialogue, exploration of alternatives, and negotiation. Conflicts are managed, even if they're rarely eliminated. But all of us can imagine, at the extreme, a community that had perfect mutual awareness, information, and empathy, and would think about conflicts of interest in radically different ways, which is perhaps why this ideal is so often part of utopian thinking.[7]

Some advances in machine intelligence have had unsavory motivations: to kill more efficiently, access pornography and drugs, or support gambling and financial greed. But more often, new forms of shared intelligence have offered an alternative to violence. Dealing with people, understanding how they think, and accepting dialogue are our alternatives to shooting, stabbing, and bombing, which are means to influence others without any need to understand them, let alone enter into collaborative intelligence.

A world of common thinking institutions, networks, and devices should also be one with less reliance on coercion, and that amplifies what is best in our nature. It should allow us to recapture a sense of possibility and progress—of "what is" as just a pale shadow of "what could be."[8]

Collective Intelligence Now

Collective intelligence is as old as civilization. But it now takes different forms. Here I give an overview of interesting recent innovations in collective intelligence.

Some are designed to observe better. Dove satellites, about the size of a shoe box, sit around 250 miles above the surface of the earth. They have shown that in Myanmar, for example, the spread of night-lights suggests slower economic growth than the World Bank's estimates. In Kenya, they count up the number of homes with metal roofs—one indicator of how fast people are moving out of poverty. In China, they have counted the numbers of trucks in factory parking lots as a proxy for industrial output. Planet Labs has helped weave together the largest network of satellites in history, constantly observing the state of the planet's ecology.[9]

Interestingly, in each of these cases, more direct pictures are replacing sophisticated representations of economic statistics. In time, these may be able to track all ships, trucks, or cars.[10]

These advances in observation can be matched in almost every other area of intelligence, from memory to analysis, and the last few decades have brought a further sharp acceleration of innovation in this respect. Half a century of advance in computation capabilities has roughly followed the forecast of Moore's law. They have given us better ways of sensing, searching, matching, calculating, playing, and killing. They include tools for logistical management, medical diagnostics, airline reservations, recommendations on what music or books to buy, navigation for drivers or walkers, speech recognition, inventory control, credit assessment, high-frequency trading, noise elimination, missile targeting, and a plethora of others.

Pattern recognition has advanced particularly fast. Facebook recognizes the people in photos posted on the social network. Google photos can recognize dogs, gravestones, and other items in pictures. Twitter's algorithms can spot pornographic pictures without any direct human involvement. Siri can interpret speech. Meanwhile, the orchestration of memory has also advanced exponentially, from databases, search engines, and linked data, to the myriad possibilities around blockchains and distributed ledgers.[11]

To categorize these new tools, new terms have sprung up, like heuristic search, logistic regression, decision and logic trees, Bayesian networks, backpropagation, convolutional neural networks, knowledge vaults, massive parallel computing projects, and recurrent neural networks. These many forms of artificial intelligence have given us vastly smarter machines for predicting, solving, and learning. Some are highly specialized. But some of the most promising ones are more general as well as more attuned to learning, such as solving complex problems by repeatedly sending back data adjusting the weightings of variables until the computer can recognize a pattern—like the shape of a hand or animal. Their algorithms learn through layers that create a hierarchy of more complex concepts out of simpler ones. Each layer provides inputs to the next one—for example, one layer looking out for edges in an image. The more layers (current technologies have advanced from a few to hundreds), the better the prospects for learning.

The most successful ones depend on huge sources of data to train machines, and have brought object and speech recognition close to human levels. Others try to mimic human abilities to make general conclusions from tiny amounts of data, abstracting from a few inputs. Meanwhile, one of the most interesting lines of development is the attempt to reverse engineer the ways in which animal and human brains work in the hope that this may lend us new insights into thought, and overcome the surprising ineptness of robots in so many tasks, from walking over uneven surfaces to tying shoelaces.[12]

The surge of thinking tools inspires and terrifies in equal measure. It promises universal, easily accessible capacity to think, but is also, in the words of Elon Musk, "summoning the demon," risking our very survival thanks to the actions of mindless engineers who simply haven't thought through the implications of their creations.

Mobilizing Human Intelligence on a Large Scale

While machine intelligence has progressed in fits and starts, a parallel movement has aimed at mobilizing human intelligence at scale, often linked through the Internet. Some aim to extract and organize knowledge (such as Wikipedia and Quora), and others aspire to manage labor (Mechanical Turk) or aggregate judgments (Digg or prediction markets). Galaxy Zoo at Oxford University in the late 2000s mobilized hundreds of thousands of volunteers to classify images of the galaxy. Foldit worked in a similar way to map proteins. In mathematics, the website Polymath encouraged people to collaborate in solving the hardest math problems, and found that many minds could often find solutions, routes to solutions, or useful new questions more successfully than mathematicians working alone.

In 2009, the US Defense Advanced Research Projects Agency launched one of the neater experiments in collective intelligence, the Red Balloon Challenge, which required competitors to track down ten weather balloons that had been tethered at random places across the United States. The winners found all balloons within nine hours, using a strategy that offered rewards to people who sighted the balloons and for recruiting friends to help with the challenge. A follow-up in 2012 required teams to find and take pictures of five people in cities in North America and Europe within twelve hours.[13] The winner, again, combined rewards for information with rewards for recruiting participants. Like disaster platforms such as Ushahidi, they tapped into strong intrinsic motivations to be helpful, but also used powerful tools to both aggregate the information as it came in and verify it. Others have tried to mobilize large groups to solve problems—from challenge and inducement prizes (like NASA's prizes for software, or Nesta's Longitude Prize, a twenty-first-century reinvention of an eighteenth-century approach to open innovation), to platforms like Kaggle, InnoCentive, and OpenIDEO. Many experiments have grappled with how to improve the accuracy with which crowds make judgments, such as by reducing the tendency to be overinfluenced by others through rewarding minority views if they later turn out to be correct.

Not all these attempts to mobilize human intelligence at scale succeed. As I will show later, their success depends on critical factors such as the modularity of problems, how easily knowledge is validated, and the incentives for participation. There are now far more examples of success, though, to draw on.

The Combination of Humans and Machines

Most of the practical examples of collective intelligence depend on combining humans and machines, organizations and networks. Just as it's hard today to imagine individual intelligence shorn of its artifacts—reading glasses or calculators—so is it useful to think of many kinds of intelligence as hybrids, combinations of people, things, and tools. We live today in the world predicted by one of the pioneers of artificial intelligence, J.C.R. Licklider, who advocated coupling humans and digital networks rather than replacing humans with machines. According to his biographer, Licklider was at the time "almost alone in his conviction that computers can become not just superfast calculating machines, but joyful machines: tools that will serve as new media of expression, inspirations to creativity and gateways to a vast world of online information."[14] This way of thinking encouraged the creation of the Advanced Research Projects Agency Network and later the Internet. It suggested that we see distributed brains as assemblies, just as the individual body is an assembly of cells, which are themselves assemblies of mitochondria, DNA, RNA, and ribosomes. It's an alternative to the vision of the Skynet of the *Terminator* films—"ahuman" systems that we can neither understand nor control—and found one of its best expressions in the theory and practice of the web, which Tim Berners-Lee described as made up of "abstract social machines" with "processes in which the people do the creative work and the machine does the administration."[15]

Many labels have been used to describe these combinations of humans and machines, such as human-machine interaction, human-computer symbiosis, computer-supported cooperative work, or social computing.[16] Licklider's premise was that the most effective intelligences will combine human and machine capability instead of simply replacing one with the other.

Much of the recent history of collective intelligence is a story of just such hybrids—combinations of human brains and computing. The spread of Google Maps is a good example. It started off with a grand ambition of organizing global geographic knowledge in a comprehensive and usable form. But Google lacked many of the crucial skills to achieve its ambition and so brought in—or to be more precise, bought in—others: Where 2 Technologies, a company founded by two Danish brothers, which provided a searchable, scrollable, and zoomable map; Keyhole, which developed the geospatial visualization software that would become Google Earth; and ZipDash, which provided real-time traffic analysis, based on information gathered anonymously from cell phone users. This assembly of different elements supplied the spine for a truly global system of geographic knowledge.

Next Google had to tap into a much wider set of skills to make the maps more useful. It did that by opening up the software—through the Google Maps API—to make it as easy as possible for other sites to integrate it.[17]

The idea was then stretched, with Google Street View, provided by Sebastian Thrun's start-up, Vutool, which was working on imaging using a fleet of cars and off-the-shelf cameras. To get Google Maps widely used, which would then help it to further attract new ideas, Google had to cut deals, including one with Apple to preload the iPhone with Google Maps as the default map app. Finally, the project mobilized the public, offering ways for people to edit and add to maps of areas they knew through Google Map Maker.

Google Maps is, in other words, less a product or service, and more an assembly of many elements that together allow the world to think in a novel way. It depends and builds on the World Wide Web, which is itself just such a hybrid assembly.[18] The Web has spawned its own ecosystem of new tools: Twitter for news, Wikipedia for knowledge, Kickstarter for investment, eBay for commerce, Wolfram|Alpha for answering questions. There are tools for finding things that are otherwise invisible (from Baby Come Home, the Chinese site created by Baidu using facial recognition software to find lost children, to BlindSquare, an app helping blind people navigate cities). There are ways of organizing knowledge (from Google's Constitute, a searchable database of the world's constitutions, to Cuba's remarkable Infomed network for doctors and public health), research (such as Zooniverse), or collective memory (such as Historypin). And there are ways of tapping many brains to make predictions, like the Iowa Electronic Markets, which try to forecast elections, or Hollywood

Stock Exchange, which predicts—quite successfully—which films will make money or win Oscars.

One of the best illustrations of a human-machine hybrid that assembles disparate elements can be found in language teaching. The time it takes to achieve rough mastery of a foreign language used to be assessed at around 130 hours, or one semester of college. The teaching program Rosetta Stone, produced by experts, reduced this to 54 hours. More recently, Duolingo combined machine and human intelligence by mobilizing 150,000 responders to test out thousands of variants of its web-based, automated language lessons. As a result, it decreased the time needed to learn a foreign language to around 34 hours—an approach that won it over 100 million learners by 2015 and has been further evolved by allowing anyone to contribute to the development of new language pairs.

Many experiments have explored how large groups can solve complex problems, working with the help of intelligent machines. The open-source movement showed that large-scale collaboration could be practical, efficient, and dynamic, and has provided most of the software for the Internet. Its ethos is parsimonious—to do programming in ways that are "lazy like a fox," in the words of one of its pioneers, Linus Torvalds—but cumulatively extraordinary. Others have used computers to manage people, such as by making computer-managed "flash teams" into problem solvers through breaking tasks down into modular elements and managing their sequencing.[19]

Some of the most interesting hybrid assemblies use platforms to aggregate and orchestrate ingenuity on ever-larger scales. Makerbot Thingiverse is a platform that hosts over 650,000 designs for the maker movement. WikiHouse shares elements of design for anyone to design their own house—encouraging users to put their own adaptations and ideas back into the commons.[20] In health, a new family of platforms allows people suffering from acute conditions to become collective researchers, turning a disparate group of patients into something more like a collective intelligence by using a mix of digital and human thought. Current examples have recruited people with Parkinson's disease and supplied them with wearable devices with accelerometers so that they can pool data about how they were faring, and a similar approach has been applied to dementia. More traditional tools use algorithms to better spot patterns and predict illnesses, like the computational pathologist (C-Path) system for breast cancer diagnoses.[21]

In business, different hybrids have emerged. A Hong Kong venture capital firm appointed an investment algorithm—VITAL—to join its board in 2014 and gave it a vote, alongside its five human board members. Baker Hostetler, a US law firm, hired artificially intelligent algorithms, an offshoot of IBM's Watson, to look after its bankruptcy cases, again with human lawyers to ensure they made sense, while another firm used artificial intelligence successfully to challenge 160,000 parking tickets.

Governments' hybrids have mainly been used to predict and prevent. Predictive algorithms predict the likelihood of a prisoner reoffending or a patient returning to a hospital. The city of New York pioneered one of the best-known examples, pooling data from five of its municipal agencies to understand the risk of fire in the city's 360,000 buildings. Around sixty factors were identified, some fairly obvious and others less so (like the presence of piles of bricks), and brought together in an algorithm that could predict which buildings were most at risk of suffering a fire. The fire service then switched its efforts more to prevention rather than cure.

All algorithms of this kind suffer from the challenge of false positives and risk optimizing for some factors and ignoring others.[22] They can easily build in biases—like the algorithms used in US criminal justice that turned out to flag black offenders far more often than white ones. Many fields are now grappling with how best to combine machine intelligence and human supervision to avoid errors of this kind. The value of an algorithm depends on both its accuracy (the probability that it will make a correct decision), and the balance between the reward from a correct decision and the penalty from a wrong one. The costs of recommending the wrong YouTube video or book on Amazon are low. But the costs of a mistake in a self-driving car or medical diagnosis are much higher, which means that there's a much greater need for human supervision.[23] A surprising effect of proliferating algorithms is a proliferation of roles for people to oversee the algorithms. Conversely, there may be more need for machine oversight of human decisions, whether they're judges or drivers. Again, the best solutions will tend to combine people and machines rather than seeing them as alternatives.[24]

A similar pattern can be seen in chess. Many chess tournaments have pitted grand masters against computers, ever since IBM's Deep Blue defeated Gary Kasparov in 1996. Deep Blue and Kasparov turned out to be fairly well matched, and in the end IBM retired Deep Blue. But a more interesting development has been the rise of players using machines to help them (in what's called "freestyle" chess), and in other cases players assisted by thousands of observers watching online and proposing moves.[25] These are illustrations of both hybrid and collective intelligence, which appear to be more effective than either purely machine intelligence or solo playing.

Perhaps the most intriguing instances of hybrids are the ones that combine humans, machines, and animals. Peru has used vultures fitted with GoPro cameras and GPS to seek out illegal garbage dumps, supported by a citizen awareness campaign and open technology that allowed the public to track the vultures, while Chad has used dogs armed with sensors to track diseases and the United Kingdom has used pigeons to monitor air pollution.

These are all examples where algorithms and sensors can amplify useful human capabilities. But an environment in which human decisions intermingle with algorithmic decisions can also amplify our uglier dispositions. Algorithms that learn fast from very large quantities of data about human behavior—tracking their clicks, purchases, and eye movements—can become adept at mobilizing unconscious desires and biases, reinforcing and taking advantage of dispositions for instant gratification and erotic stimulus. The spread of fake news is one example; others include the manipulation of gamblers' optimism or shoppers' hunger for what's shiny and new. The result risks leaving us with a cartoon caricature of human nature and encouragement of habits people would probably prefer to rein in than to amplify.

The Trade-offs of Intelligence at Scale

The best examples are pointers to a future of radically enhanced capacities to think that merge human and machine brains (and occasionally other species too).[26] But most of the successful examples of recent years have addressed fairly discrete tasks, tasks where there is little disagreement about their nature, and knowledgeable communities.

Unfortunately, many of our most pressing tasks aren't like that. They involve fuzzy definitions, conflicting interests, and less clarity about whether the answers are right (only time tells). The problems can't so easily be distributed because some of the organizations with the best skills don't want to offer them up or may feel threatened by the likely solutions.

Much has been learned recently about the subtleties of real-life collective intelligence that doesn't quite fit the expectations of the pioneers. The first generation of crowdsourcing devices often misunderstood how problems are solved. Rather than discrete answers solving discrete problems, there's usually a much more iterative process of problem definition and solution development (I discuss this in more detail in chapter 12). We circle around problems and probe them before solutions emerge.

Another important lesson is that negative network effects can be as strong as positive ones: the crowd may generate more noise than signal, let alone wisdom, and having too many people involved in solving problems can impede rather than help progress. Similarly, although social networks are good for quickly circulating information or finding answers to certain types of questions, they may not increase social learning—the ability to answer future problems or questions more effectively.[27] A bigger community can be helpful where more data and information are needed, to generate new options, or where the parameters are clearly defined—like voting in an election or solving a well-defined problem. But they may be unhelpful where the selection criteria are fuzzy and a lot of subtle information is needed to make judgments.[28]

Assemblies

It's appealing to think that a single organizing model could guide a collective intelligence—the beauty of markets, a civil service made up of the most brilliant minds, or a science system of peer criticism. As will become clear through the course of this book, however, most successful collective intelligences look much more like hybrids, assemblies of multiple elements. Google Maps is a typical example of a mashed-up hybrid—a new kind of "social machine" in which the human and machine are seamlessly interwoven.[29] Many large companies have comparably complex assemblies. Amazon, for instance, combines its "comprehensive collaborative filtering" engine to generate recommendations based on the choices made by other people, tools like one-click to remove friction (and second thoughts), "anticipatory shipping" to send goods to local distribution centers in advance of people making purchases, graph theory to optimize delivery routes, price optimization to fit prices to the customer and local market conditions, and many others.

To work well, and serve a whole system, whether that's within a company or run as a public good, an assembly needs to combine many elements: rich sources of observation and data; models that can make predictions; capacities to interpret and analyze; abilities to create and innovate in response to new problems and opportunities; a structured memory, including of what's worked in the past; and a link into action and learning that's aligned with how people really behave. The test of these elements, when linked up, is then whether they help a whole system think and act more effectively.

For now, there are only pointers to the kind of assembles that might be possible in the future. Some focus on the environment. The Planetary Skin was set up by NASA and Cisco as a global nonprofit research and development organization that would survey the state of the world's ecological systems to help people better prepare for extreme weather events, or problems caused by shortages of water, energy, and food. Hewlett-Packard's Central Nervous System for the Earth is a parallel project, with equally bold ambitions. Both struggled with finance. Europe's Copernicus program—which has a similar goal of mapping the state of Europe's ecosystems—may prove more successful, in part because of a more solid funding base.

Other ambitious projects are being developed in medicine, including, for instance, MetaSub, which maps the global urban microbial genome so as to better understand patterns of antimicrobial resistance.[30] The creation of AIME, a global network using artificial intelligence to track and predict outbreaks of Zika and dengue, is another good example—combining sophisticated observation, computing power, and clever behavioral incentives (such as learning from Pokémon GO how to reward the public for hunting out breeding sites).[31]

One of the most comprehensive existing collective intelligence assemblies quietly supports cancer treatment in England's National Health Service through the National Cancer Registration and Analysis Service. Despite the mundane name, this is an extraordinary feat of organization that points to how many public services and whole systems could be run in the future. It links thousands of records—including the three hundred thousand new cases of cancer in England each year. It brings together diagnoses, scans, images, and past treatments. The data feed into predictive tools to help patients choose different treatment options. Where necessary, the data are linked to genetic information, or other data sets that help to predict if the illness could lead to debt or depression, and market research information to help better target public health messages. The whole array of information is then used to guide the day-to-day decisions of doctors and increasingly patients too. Unusually, it has the advantage of a simple funding base (part of the nearly $10 billion annual budget for cancer care), and its value is obvious, including to patients who can access the data themselves.

Global medicine is likely to come closest to a comprehensive assembly combining data gathering, interpretation, experiment, and the systematic organization of memory (and potentially in the near future can link these into networks of sensors and implants as well as drones gathering data and distributing drugs and samples). Medicine is helped by its relative wealth, high status, and global nature. But all these assemblies struggle with economics: Whose job is it to pay for these things?[32] They can work well within individual businesses (like Amazon) or funded by a hugely profitable firm (like Google). In principle they could be funded as clubs, with small payments from many users. But without at least some funding from governments and taxpayers, many look set to struggle.

They also find it hard to link observation to action, and become part of the daily work of the professions and daily life of the citizens they're designed to serve. That requires the information they produce to be accessible, relevant, and timely. Assemblies are in part technical designs, but they only become useful if they connect to action, which requires them to be sophisticated about behaviors, cultures, and organizational norms, all of which may be more taxing than the design of sensing systems and algorithms. They are beginning to show how at a global level, though, a wide range of resources—from satellite networks to university labs, public health officials to teachers—could be linked together into something more like a single brain and truly global nervous system.

These compound examples have parallels in the natural world. The eukaryotic cell evolved as a mashed-up, crashed hybrid of bacteria converging to create something new. Our own language has its own parallels too. The verb "to be" in English is a good illustration, as a mashed-up, crashed hybrid of at least four different word groups (be, am, is, were . . .) that through evolution have been combined together. Or think too of the modern computer with its sensors, CPU, keyboard, and connections. Our most useful tools are often amalgams and assemblies, mongrels and hybrids, and none the worse for that. They evolve through shuffling the elements—trying out new combinations in iterative ways.

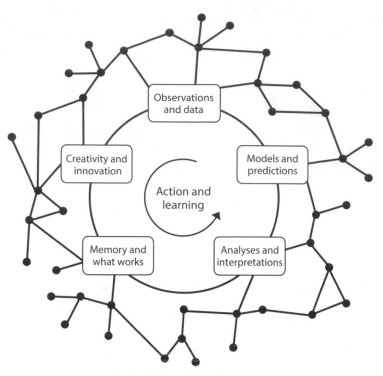

Figure 1. Assemblies

But such assemblies are costly. They require work, investment, special-ized skills, and machine intelligence. And for our most pressing problems, there is rarely the appetite or wealth to invest for creating them. As a result, our shared intelligence consistently underperforms, dots are not joined up, patterns are not recognized or acted on, and lessons are not learned.

COLLECTIVE INTELLIGENCE, POSSIBILITY, AND ANXIETY

Before concluding this first overview of collective intelligence today, I want to briefly mention an odd feature of all kinds of collective intelligence. Op-timists would like to believe that collective intelligence simply enhances our ability to master the world and solve problems. Instead, what has been discovered again and again is that it has a contradictory effect on mood and psychology. On the one hand, it expands our shared sense of what is possible. Novel utopias and imagined societies come into view, inspiring a fresh sense of the plasticity of the world. At the same time, however, enhanced collective intelligence also makes us more aware of precariousness and risk. It brings, in other words, both hope and anxiety.

This has been the experience of the last half century in health—where greater understanding of how to improve health, prevent diseases, and enhance fitness has come with greater anxiety, not least among the "worried well." In the same way, we have an ever-greater ability to shape and manage environments, but also an ever-greater awareness of the fragility of systems. In short, an evolution that takes humanity to ever-greater levels of collective intelligence will not be a comforting evolution or a path that promises an end to fear. It instead will take us to new levels of awareness that will expand horizons and uncertainty at the same time, including our anxieties not just about machines that outclass us but also about what we do if the machines fail, if the network crashes or electricity turns off, and we no longer have the reserves of human skill to cope without them.

PART II

Making Sense of Collective
Intelligence as Choice

HAVING DESCRIBED THE RAPIDLY EVOLVING PRACTICE of collective intelligence, I turn in part II to the tools, concepts, and theories we need to make sense of how intelligence works at large scale.

I show the elements that support it—from observation and memory to judgment—and how they're brought together to interpret and decide. A crucial step in the argument shows the loops we use to learn, and how when we run into barriers, we can create new categories or new ways of thinking.

I explore the traps we can fall into, whether because of deliberate sabotage, conflict, or self-deception, and how to fight against the many enemies of collective intelligence. I demonstrate that more doesn't always mean better; more data and more participants should make for more intelligent decisions, but this is by no means guaranteed.

I then describe how a sense of a "we" can come into existence—a collective subject that could be a group, organization, or nation, while also showing the limits of that idea.

By the end of part II, the reader should have an armory of ways to look at any organization, city, or field, and make sense of how well it thinks as well as how it could think and act more effectively.

- 3 -

The Functional Elements of Collective Intelligence

THE ASSEMBLIES OF COLLECTIVE INTELLIGENCE bring together distinct capabilities—for seeing, analyzing, remembering, and creating—so as to make more effective action possible. Within organizations or across societies, these turn out to require different approaches, cultures, and organizational methods.[1] The kinds of people who are good at observation may be poor at creativity, and vice versa. Memory is organized in different ways than prediction.

Each element of intelligence also requires energy and time, which means that there are trade-offs between them.[2] More of one means less of the others.[3] It's possible that in a few decades, the world will be convinced that there is a *general* property of collective intelligence—an IQ for groups. But for now it looks highly unlikely.[4]

So what are the distinct elements that combine together to allow for thought and action on a large scale? Here I describe the main ones.

A live model of the world. The end point of intelligence is action in the world. As Johann Wolfgang von Goethe puts it in *Faust*, "Im Anfang war die Tat" (in the beginning was the deed). But to be able to act, we need more than motives, and more than data or inputs. We need an implicit or explicit model of the world—how it works, how things cause other things to happen, how other people behave, and what matters. Without this, all inputs are meaningless. Intelligence constructs its own internal environments and then tries to align them with the external environment. This is the stance of Bayesian thinking—we begin with a "prior" and then see whether the data confirm it, and so we steadily improve the probability of that knowledge being correct. The model is live and hungry; it is eager to predict and test. It has to work out causes and then adjust when the world fails to obey our predictions.[5]

Using models, we can think things through rather than having to do them to learn. Thought replaces or complements trial and error. This is a capacity that Daniel Dennett describes as Popperian—the hypothesis dies instead of us by precorrecting errors. It's not unique to humans. A squirrel contemplating a jump from one branch to another tenses its muscles as it simulates the jump, before deciding that, after all, the distance is too far.[6]

Our models need to be useful, and the intuitive physics and psychology we are born with, and our sense of number and space, are useful. But for humans and groups, though probably not for machines, they also need to be coherent.[7] Our models don't all have to add up, and the mark of a fine mind is the ability to hold two opposing ideas at the same time. Yet we struggle if our models are too divergent and would rather ignore facts that are too challenging if we possibly can.[8]

Observation. Next comes the ability to see, hear, smell, and touch the world, feeding into our existing models. By some accounts, 40 percent of human brain activity is associated with seeing. No living intelligence can survive without an ability to observe the surrounding environment. Animals vary greatly in their ability to observe, and humans are not particularly impressive. We hear less than dolphins or dogs; see less than many birds, some of which can see ultraviolet; smell less than some insects; and sense less than bats. Our perception of time has a lower limit of around a tenth of a second (which makes us far slower than, for example, trading technologies making decisions at a speed of well under ten microseconds).

But a high proportion of learning comes from observation. We are a species of copiers. It's deep in our makeup, designed into mirror neurons and habits of copying others frequently below the surface of consciousness that are apparent in children within the first hour after birth, and then aided by social interaction, conversation, and media. A surprising proportion of advances in science began just with seeing things in a new way—thanks to microscopes and telescopes, or statistics. The achromatic-lens microscope in the early nineteenth century paved the way for germ theory, and X-ray crystallography in the early twentieth century played a vital role in the later discovery of the structure of DNA. In the same way flows of data—for example, about how people move around a city, or how blood cells change—can prompt new insights.

Observation is not a raw capability, though. What we observe are not unvarnished sensory inputs but rather are framed by models, expectations, and our views of how the world works and what matters. One of the surprising findings of recent science is that more visual signals leave the brain than enter it, expectations of what is to be seen that are then either corroborated or disproved by the data coming from the eyes. What we see depends on what we know as well as vice versa. That is why we are so good at ignoring unexpected or inconvenient facts, why experts can be stupider than amateurs, and why theory-induced blindness is such a common failing in sophisticated societies.

Attention and focus. The combination of models and observations—or inputs—then helps intelligence to focus in order to distinguish what matters from what doesn't. Pablo Picasso once commented that "there are so many realities that in trying to render all of them visible, one ends up in the dark." So we learn to select.

The various traditions of mindfulness teach how to attend to attention itself—to be aware of the ways in which minds wander and thoughts crowd in. The ability to focus well is recognized as a particularly useful trait, whether for individuals or groups. Walter Mischel's famous experiments with children resisting marshmallows showed that a child who could resist for fifteen minutes would end up with a test score far higher than one who could only resist for thirty seconds. He found that the critical ability was being able to look away, to resist temptation by ignoring the marshmallow—a metaphor for disciplined focus.

This ability to concentrate is not only a crucial attribute of personal effectiveness. It is also essential for organizations to have the ability to "kill puppies," to stop the projects that are interesting and appealing yet ultimately a diversion.

There's an intriguing parallel in vision. Visible light is made up of all the colors of the spectrum. When visible light passes through a lens, blue and violet light refracts more than orange and red, meaning that each color focuses at a slightly different point, creating a slight blur around the edge of an image. Professional shooters use lenses that reduce the amount of shorter-wavelength blue light with a filter to improve clarity. By seeing less, they see more. Yet focus can also be a weakness. Too much focus and too little peripheral vision condemns you to fail to spot the novel pattern, the unforeseen threat.

Many highly intelligent people, machines, and organizations struggle to concentrate, and suffer from the vice of selection failure: they drown in data, get lost in too many options, and don't know how to judge or decide. For them, quantity defeats quality and noise overcomes meaning. But all of us suffer a version of this vice, because we can only ever know in retrospect which information should have been attended to, and which should have been ignored.

Analysis and reasoning is the ability, or abilities, to think, calculate, and interpret. This is the main territory for algorithmic intelligence. It involves step-by-step processing and analysis along with options, thinking things through, and working out what causes what. There is a vast literature on the many ways in which humans reason. We can reason through deductive inferences from premises, and then see when data contradicts or confirms our models; through induction; or through analogy and formal logic. We can reason through axiomatic theory and abstraction, as happens in physics and mathematics, or we can reason through classifications linked to observation, as in chemistry. We can learn to understand causes as either direct or probabilistic, and distinguish systemic causes from predisposing and precipitating ones. Some of our reasoning capacities are strong, such as when spotting patterns or making sense of social relationships. But others are weak, and certainly much weaker than computers, notably our capacities for calculation and matching.

Perhaps the core of any intelligence is this ability to use the combination of models and observations to then analyze and interpret. What caused what? What effect did my reassuring words to my lover have? Did I bring up my children the right way? Was this the right apartment to have rented? In computer science, this is sometimes called the "credit assignment" problem of intelligence: How do I decide which actions led to good outcomes if there are long delays or multiple factors involved (and in most real situations, there are likely to be many more factors than we can know)? Each time we make a decision, we have to make a judgment about causation. We also have to ask whether the rules and models that applied to a previous situation now apply to this one. Hopefully we update our models when they no longer work, although the vagaries of human psychology mean that we can cling desperately to failed models when the world refuses to obey them.

Creation or creativity is the ability to imagine and design new things. This was once thought to be a uniquely human capability, though of course we can't easily know just how creative a dolphin or eagle really is, and distinct cultures have been observed in chimpanzee groups. Our brains are well placed to imagine, and we show from an early age the ability to make new things, sometimes using analogy, metaphor, or combination, and throughout life creativity is more playful than the other elements of intelligence (where play is not so much the opposite of work as the opposite of boredom).

Creativity plays its part in every domain, from daily life to science. For example, when August Kekulé saw a benzene molecule in a daydream as a snake eating itself, he realized that the molecule was shaped as a ring, just as when James Lovelock saw the earth as a creature preserving itself, he found an insight into how climate might function. It's common for the unconscious to play an important role in creativity—solving problems and seeing patterns that the conscious mind struggles with. In its more extreme forms, creativity gives us the vivid breakthroughs of prophecy that lie at the heart of many religions and art forms.

Motor coordination is the ability to act in the physical world—to connect a thought or observation to the movement of a hand or leg. The words dexterity, management, and digital all reflect the importance of these abilities to link thought and action, to know when to run or fight. Groups and organizations have parallel capacities to orchestrate the physical movements of energy and armies, trucks and trains, and with the rise of the Internet of things, the physical world is becoming ever more integrated with other types of intelligence.

Memory is the ability to remember, both short and long term. Human thought often appears feeble because of a working memory that can only handle four items at any one time, which is why we depend so much on pen and paper, or our digital devices, for even simple tasks like making a list or doing a calculation.[9] But we can remember a vast amount, we learn repertoires of action that draw on experience, and our memories and monitoring of ourselves both influence evolving models of the world. Any sense of a coherent self depends on memory. Yet memories fade with quite-predictable patterns of forgetting, and some memory may be suppressed; organizations and groups, like individuals, may be shaped as much by the memories they have deliberately forgotten as by the ones they can readily access. For organizations as for individuals, the challenge isn't just to remember; it's also to retrieve the right memory at the right time.[10]

Empathy is the ability to understand the world from another's perspective. Atticus Finch in *To Kill a Mockingbird* said that you never really understand a person "until you climb into his skin and walk around in it." But that isn't quite true. We can empathize, and the roots of empathy are innate. Empathy grows through practice and example. It's helped by observation—the subtle signals from facial expressions, bodily pose, or tones of voice—but also depends on interpretation and sympathy—the ability to feel with another and not just analyze his or her feelings. Empathy takes us toward love, the deep feeling for another that connects us to the infinite in ways that transcend the rational mind. Empathy can also fuel harmful collective behaviors when it overrides intelligence, though. All too often, groups turn heightened empathy for one of their own who has become a victim of violence into hatred of another group.

Judgment is the ability to make decisions, and leads to wisdom along with the ability to make sense of complexity and integrate moral perspectives. Judgment is where we pull together analysis and reasoning with experience and intuition. It is partly rational but also partly emotional. Without emotions to guide us, we struggle to decide in light of scarce and contradictory information. Big minds have to be matched with big hearts, whether our interest is with individuals or groups.

Wisdom is the ultimate kind of judgment. It tends to be more contextual and less universal than other kinds of reasoning. It integrates ethics and attends to appropriateness. This is the perhaps-surprising finding of all serious study of wisdom and judgment. What we recognize as the highest intelligence is not the application of standardized protocols to many different types of problems but rather the ability to understand the specific character of particular places, peoples, and times. Science in this sense can be at odds with wisdom. It has come to mean the quantitative study of large sets (stars, atoms, or cells) seen from afar. Wisdom, by contrast, involves the more qualitative understanding of things seen close up or lived. We associate it with an ability to take a long view, and to allow things time to mature, gestate, and evolve, even through periods of incoherence and failure.

INTELLIGENCE AND VERIFICATION

The relationship between wisdom and other aspects of intelligence has fascinated philosophers for several millennia. Aristotle distinguished three different ways of thinking. *Episteme* is the logical thinking that applies rules, and *techne* is the practical knowledge of things, while *phronesis* is wisdom. Each has its own logic of verification. Episteme can be verified through logic or formal experiments. It only takes one counter example to disprove a rule or hypothesis. Techne is tested by practice: Does something work or not? Phronesis, on the other hand, is determined by context, and can only be verified through applying it to choices and learning step by step whether decisions really do turn out to be wise or not.

In some accounts, this leads to an ideal of action as shaped by seeking out energies and possibilities, and then going with them, as opposed to following the linear logic of purposes, strategies, and actions. For the wise person, success ripens, emerges, or matures rather than following a neat line of causation.

The Elements of Collective Intelligence

A live model of the world
Observation
Focus
Memory
Empathy
Motor coordination
Creativity
Judgment
Wisdom

How Technologies Enhance the Functional Capabilities of Intelligence

It will immediately be obvious that computers can effectively perform many of the varied capabilities of intelligence described above—from observation to memory. Computers have vastly enhanced our ability to see, calculate, and remember, and their capacities have far outstripped our ability to keep up and made us godlike in our powers to observe.

Motor coordination has been transformed to an almost-comparable extent. The combination of sensors, mobile communications, and computing makes it possible to organize many physical phenomena in radically different ways—notably in energy and transport—with machine-to-machine communication replacing the need for human agency.

Computers are effective at storing and retrieving items, such as a name or image. Human memory is more fluid, dependent on selective strengthening, combining, or weakening, and so we forget much of what we know or see, and can all too easily remember things that never happened. Looking to the future, blockchain technologies have the potential to further transform collective memory, providing a shared record of actions and transactions in ways that are secure but also transparent, such as by tracking every diamond in terms of its origins and ownership so as to cut off the rewards for diamonds stolen in conflicts, or tracking the origins of every product on sale in shops to guide consumers concerned about the environment or exploitation of workers.[11]

Computers can increasingly read emotional dispositions from facial movements or voice stress. This isn't full empathy in a human sense, but it successfully mimics and amplifies some aspects (when, for example, sensors can read the emotional response of a whole audience). As I will show in chapter 11, it may not be long before we use computers to manage and facilitate meetings, sensing moods and offering more strategies than the typical human chair or facilitator can.

Computers can also capture complex patterns of experience and augment decisions. "Building information modeling" programs are a good illustration; they offer three-dimensional models before and during the construction of a building, providing detailed representations of designs, specifications, cost estimates, and most crucially, "clash detection," showing when pipes are likely to hit each other or a plan might be incompatible with a local planning rule. Here technology captures a great deal of formal and informal professional knowledge, but also helps to focus human attention on more creative work.

Other elements of intelligence are, by contrast, far less touched by technology. Creativity remains in the balance. The thousands of experiments with computer-generated art or music are intriguing rather than persuasive. Computers can generate plausible and appealing music in the style of Bach, soundtracks and jingles, or haiku poems, and can now combine the representational content of a photograph to the style of a particular painter.[12] It's unwise to claim that there is any inherent reason why computers cannot in time excel and potentially create entirely new art forms.

Capacities to judge are less touched by computing. Diagnostic artificial intelligence is becoming better at judging where there are large data sets, but poor in conditions of radical uncertainty. Although great progress has been made in observation, much less progress has been made in explanation, or the machine equivalent of "theory of mind"—the ability to empathize and see things through another's eyes.

Our relationship to these tools is complex. Our abilities have been massively amplified, and will become ever more so. We are learning to use these tools to augment, challenge, and corroborate our own senses, evolving fairly fast into hybrid organisms that combine machine and wet matter. Already tools like Cortana's artificial intelligence personal assistant or Google Now can guide us to avoid unnecessary mistakes—forgetting a crucial anniversary or messing up our body chemistry.

But we have also learned that every tool that amplifies and orchestrates human intelligence can become a trap. Selecting the data that fit a particular task can lead us to rely too heavily on those data, and miss more important data that at first appear peripheral. Sophisticated tools for tracking performance have repeatedly turned out to entrap, as organizations hit the target but miss the point. Predictive tools that make recommendations based on our past behavior can turn us into caricatures of ourselves rather than helping us to learn. Intelligent devices can quickly evolve from tools into masters that either implicitly or explicitly shape us in their image rather than the other way around, or offer back to us a distorted mirror of our own often-confused selves. This is why we have to learn both how to use digital tools and when to reject them so that we don't end up trapped in new cages of our own making.

It's an interesting thought experiment to imagine a computer designed for wisdom or phronesis, which would adapt its algorithms to the specificities of context and perhaps avoid some of these traps.[13] Yet these would be quite different from the machines we have today, ancestors of Turing, which are rule-based machines, pure expressions of episteme. It helps that we're learning more about how human expert intelligence works in practice, with intuition, heuristics, and emotions allowing for more efficient judgments than linear reasoning on its own.[14] This may take us to the ultimate human-machine hybrids that combine humans who have learned to be more machinelike with machines that have learned to think more like humans.

Balance and Imbalance

In our daily lives, we have to strike a balance between the different capabilities of intelligence. It can be problematic to live too much in memory, or be too analytic, too creative, or too judgmental. Indeed, minds that become too fixated on just one element suffer and at the extreme become ill (like being unable to avoid memories or so rational that you struggle to make decisions). Achieving the right balance between all the different elements of intelligence is also critical for any group or organization, and may be harder if technologies massively amplify just one. Too much memory, and you risk being trapped in the past. Too much reason, and you risk being blind to intuition and emotion. Too much creativity, and you may never act or learn. Too much coherence, and you may not see when your methods no longer work.

I've repeatedly experienced examples of imbalance. For instance, within governments there are departments where memory is so strong that every new possibility is discounted on the grounds that it has been tried before. I also found the opposite case in other governments, where there was no organized memory of the last time something similar was achieved, thereby confirming George Santayana's famous comment that "those who fail to learn from history are doomed to repeat it." Some organizations invest heavily in observation, but much less in the ability to interpret, driving them into perpetual paranoia. Stalin's USSR was a good illustration of this, and in a different way, some contemporary firms become hyperresponsive to day-to-day data, but lose sense of the bigger picture. As William Binney, a former technical director of the National Security Agency, told a British parliamentary committee in January 2016, the bulk collection of communications data was 99 percent useless and "cost lives . . . because it inundates analysts with too much data."

Creativity is a virtue, but organizations can become too creative, unduly focused on what's new and interesting. By failing to learn from or remember the achievements of others, they continuously reinvent the wheel.

For the individual, the methods of mindfulness help—amplifying each of these capabilities, so that we observe more acutely or create more imaginatively, seeing our thoughts for what they are. For a group, too, mindfulness is a meaningful concept—and may require some hierarchy, or some means for the group to decide what to concentrate on and how to shift resources between the different components of intelligence. Perhaps a meta-intelligence is not a general intelligence in the Turing sense but rather an ability to switch between different types of intelligence as appropriate. This requires some other anchoring—in a task or mission, or an identity, since it is only with reference to something outside the intelligence that judgments can be made about how to organize the intelligence itself.[15]

The importance of balance also shows up in a rather different way in relation to complex thinking. J. Rogers Hollingsworth's detailed analysis of hundreds of creative and complex scientific breakthroughs attributed them to some obvious factors like access to resources, but also less obvious ones like scientific diversity, good connections to widely spread networks, and leaders with a sense of how to integrate different fields. The most successful scientists, he found, were often deeply involved in fields far from science—such as music or religion—which helped them see novel patterns. But the research also showed an inverted U for the relationship between diversity and results: too much diversity led to noise rather than breakthroughs, just as too much communication presumably squeezed out time for reflection and novel thought. The best scientists, in other words, knew how to get the balance right, even though that was hard to plan since their minds tended "to evolve in an unplanned, chaotic, somewhat random process involving a considerable amount of chance, luck and contingency."[16]

The Dimensionality of Intelligence

As we've seen, all exercises of intelligence, whether individual or collective, have similar underlying structures. They combine a set of observations, data, and other inputs of information, which may be either partial or comprehensive; some ways of interpreting and analyzing those that data, using preexisting or new models; a decision and action; and then some means of adjusting in light of what happens, which can be called either feedback or learning.

The relatively simple cases are ones where the data are unambiguous, the interpretative model is stable and established, and the feedback is quick. A lizard catching a fly, person driving a car, or missile attacking a target would all fit neatly into this pattern. Much of the advance of artificial and machine intelligence has involved expanding the quantity of data gathered, improving the interpretative algorithms, and adjusting models in light of immediate feedback. This is why they are so well suited to playing games like chess, Jeopardy, or Go.

Some choices are binary, linear, and rather like computer programs (Do I buy this or that milk? Take this or that train?). Artificial intelligence tends to do best in artificial situations—like playing games or analyzing streams of well-ordered data—and worst in more multidimensional situations.

The challenges come when there is neither sufficient data nor sufficiently stable and reliable interpretative frames to make a decision. Much of what we recognize as useful intelligence has to be able to handle incommensurable quantities, deal with considerations that are different in nature from each other, and synthesize decisions out of messy elements rather than determining them in a linear way. Having made a decision, it may take years for us to discover whether the choice was right or not. Most of the really important choices fall into these categories, where the data, models, and feedback loops are fuzzy, such as choosing a life partner, job, or whether to move to another country. We can't pin down the risks and probabilities, and so we have to make decisions clouded by uncertainty.

To help us, we circle around the problem, looking at it through multiple lenses, rational and emotional, imaginative and analytic, to get a feel for the right decision as opposed to addressing it in a linear way with a single logic. Groups are the same. They try out many different ways of thinking about the problem, using argument and dialectical thinking, instead of either induction or deduction. They may then widen the range of choices on offer—a creative act generating options that can't be drawn solely from existing data. Then they may apply criteria, weigh the lists of pros and cons (a method used for five centuries or more), or listen to their subconscious.

What are the implications? A general theory of collective intelligence needs to address the *dimensionality* of choices. In statistics, this refers to the number of variables involved.[17] But just as important are cognitive dimensionality (how many different ways of thinking, disciplines, or models are necessary to understand the choice), its social dimensionality (how many people or organizations have some power or influence over the decision, and how much are they in conflict with each other), and its temporal dimensionality (how long are the feedback loops). The more dimensional

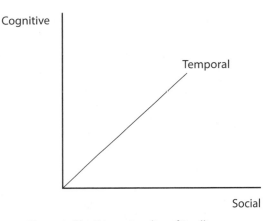

Figure 2. The Dimensionality of Intelligence

the choice, the more work has to be done, and the more resources have to be expended to arrive at a good choice.

Any choice can be mapped in this three-dimensional space. The majority of collective intelligence experiments are close to where the axes meet, such as offering proposals for a mathematics question or advice to a chess player. There is one primary cognitive frame for answering the question (however complex), decisions rest with a single individual or organization, and there are rapid methods of verification (Is the game won or not? Does the proof stand up?).

Current computers are much better designed for low-dimensional problems than high-dimensional ones. But there is no inherent reason why they cannot shape much more complex questions in time. They can offer multiple interpretative frames to a group, individual, or committee grappling with a question (rather as existing deep learning tools propose new algorithms in response to training data). They can be used to map social patterns of power, communication, and influence, again helping human decision makers not to ignore important factors. And they can attempt simulations of the results of decisions.

Many organizations tackle high-dimension problems with low-dimension tools. If you're powerful enough, you don't need to bother with involving other, less powerful organizations in making decisions. If your discipline has sufficient legitimacy, it can ignore the other disciplines that might have something to say. Yet the end results are worse.

- 4 -

The Infrastructures That Support Collective Intelligence

FOR A GROUP TO THINK WELL, IT IS NOT ENOUGH to have many clever people and smart machines. Instead, large-scale thought depends on infrastructures—underlying systems of support, both physical and virtual. Past attempts to organize collaborative thought on a large scale show us how these can be organized well.

Richard Chenevix Trench's plan to create the *Oxford English Dictionary*, or *OED*, was an extraordinary example of a truly collective endeavor. The dictionary's purpose was to find both the meanings of every word in the English language and the history of those meanings. Previous dictionaries had offered definitions, but these reflected the opinions of their authors. The *OED* attempted to be empirical, scientific, and comprehensive. The project therefore required that someone should read all available literature and gather every use of every word. This was clearly an impossible task— certainly for a single individual or small group.

Trench recognized this, and at a meeting at the London Library in 1857, argued that this task was so obviously beyond the ability of any one person that it must instead be "the combined action of many."[1] He proposed recruiting a huge team of unpaid volunteers to map their own language and heritage. A collaborator, Frederick Furnivall, duly assembled just such an army. Each volunteer was asked to select a period of history, read as much as they could from that period, and compile word lists. They were told to search for words of particular interest to the dictionary team, using slips of paper that contained the target word, the title of the book, its date, volume, and page number, and the sentence using the word.

The principles of many later ventures of this kind were established here: a compelling goal; a large army of volunteer contributors; strict rules to govern how information was to be organized and shared, including what we would now call metadata; and a central group of guardians to keep the whole exercise on course.

Herbert Coleridge became the first editor of what came to be called *A New English Dictionary on Historical Principles*. He expected to receive around a hundred thousand slips, and that it would take at most a couple of years or less, "were it not for the dilatoriness of many contributors." Instead, over six million slips of paper came in, and it took over twenty years before he had the first, salable part of the dictionary. As the author of a history of the exercise wrote, "It was this kind of woefully naive underestimate—of work, of time, of money—that at first so hindered the dictionary's advance. No one had a clue what they were up against: they were marching blindfolded through molasses." March they did, however, and their project succeeded—a model for many more recent ones, like Wikipedia, that orchestrate intelligence on a large scale.

This example showed how, with the right orchestration, the properties of a group could far exceed the capabilities of any one part. Compare a brain and neuron, an ant colony relative to an ant, or a society relative to its members. The larger-scale intelligence isn't just smarter in terms of quantity; it's also radically different in terms of the quality of its thought. So what can we learn from this instance? The *OED* shows the crucial tools that underpin collective intelligence—the common infrastructures that allow common thought to flourish.

Common Rules, Standards, and Structures

The first vital infrastructure is a set of agreed-upon rules and standards, like the classifications that were so vital to the dictionary. These are a necessary condition for knowledge commons of any kind, and greatly reduce the transaction costs of thought and coordination.

The *OED* wasn't the first dictionary. It came long after Denis Diderot's 1751 *Encyclopédie*, Ephraim Chambers's *Cyclopedia*, and Samuel Johnson's 1755 *Dictionary of the English Language*. But it better exemplified the idea that codes and classifications could unlock the world.

Our ancestors depended on the detailed lessons about how to survive, and use plants, bones, or skins that together constituted a culture that made it possible to survive in deserts or arctic tundra. Both individual and collective intelligence evolved in response to challenge, as a turbulent climate gave people no choice but to adapt to new plants and animals, through a roller coaster of heat and cold that rewarded the fluid, flexible intelligence of the human brain. But that intelligence could only be marshaled and passed down through the generations if it could be turned into structures—memorable lists, taxonomies, clusters of relationships, or the rules of prosody governing words in song and myth—and these are at the heart of preliterate cultures.

Much of the more recent history of literate civilization can likewise be read as an evolution of more powerful and standardized tools for thinking: libraries to hold memory as well as universities to both remember and create along with firms and markets to coordinate production, parliaments to make decisions, hospitals to cure, academic disciplines to organize deep knowledge, and artists' studios to push the boundaries of creativity. More recently, Google began with the audacious goal of copying the whole World Wide Web in order to represent and then search it.

These in their different ways orchestrated society's abilities to observe, interpret, remember, and create, putting them into dedicated institutions and buildings while training cadres of professionals to concentrate, organize, distribute, and use new knowledge. What held them together were common categories and classifications. Carl Linnaeus's botanical taxonomic methods mapped the world of plants and could then be applied to human social phenomena including the production of racial taxonomies. Dmitri Mendeleev invented the periodic table (reputedly arising out of a game of Patience). Other formal systems, closer to the world of the *OED*, included Jacques Charles Brunet's Paris Bookseller's classification (1842) and the Dewey Decimal System (1876). All were ventures in epistemology—defining what counts as knowledge, what matters, and the relationships between things.

These ways of thinking together spread to every field. In business, innovations in management accounting were needed to handle the greater complexity of nationwide railway networks, complicated steel plants, or chemicals. Without new ways to see not just the physical movements of things but also the intangible value moving through the system, there was no way to keep the enterprise profitable.

In the world of data, the debate about common standards that make thought easier adopts the language of organized religion: "canonical rules" to make data easier to manage and manipulate, and shared "registries." Without standardized metadata and tags, it's hard to organize information to be useful—yet these are still missing in many important fields.

Standards can be almost magical. One of the inventors of the spreadsheet, Bob Frankston, described it as "magic sheet of electronic paper that can perform calculations and recalculations, which allows the user to solve the problems using familiar tools and concepts"—an approach that led through to Excel and others, and coming equivalents in big data. The spreadsheet is a fascinating example of both the power and risks of standardized tools for collective intelligence. It was hugely influential and useful in allowing large numbers of people to collaborate on complex financial representations, but also hugely risky when there were insufficient checks on or records of who had changed what number, all amplified by a natural tendency to confuse the representation for reality.

One Finnish prime minister even went so far as to warn against the risks of governing as if by Excel. But governments, too, have learned to create infrastructures, standard classifications, and tools for thinking. The rise of statistics was driven by governments' need to see and control, and in particular their need to raise revenue. These in turn made possible the development of probability theories that uncovered surprising findings, like the constancy of birth and crime rates, or rates of suicide and insanity, or the numbers of immigrants and emigrants, suggesting constant laws amid the apparent chaos of human life. Modern social science emerged from the resulting social physics and "political arithmetic," helped by curious governments. The UK Parliament's Blue Books collected data on a vast scale. New categorization and visualization systems devised by figures like William Playfair, inventor of the line graph, bar chart, and pie chart, were powerful tools that helped people think in novel ways.

This hunger for comparisons fueled what the historian Ian Hacking called an "avalanche of numbers" in the early nineteenth century. Today's computational social science is having a comparable impact, with another avalanche of numbers and tools requiring governments to collaborate on everything from predictive algorithms combining data from many departments to dynamic visualizations of complex data.

Seen in the long view, standardized measurements have become both more pervasive and more diverse. They have evolved from taxable things (buildings, animals, and people) for government to measure society, to measures for society to judge itself and what government is doing (such as the Organization for Economic Cooperation and Development's PISA scores for school exam results, or the measures designed by Transparency International, an NGO that ranks countries according to perceived levels of corruption). They have evolved from physical objects (such as steel production) through aggregate concepts (like GDP and GNP) to intangibles (such as innovation indexes or measures of the value of creative industries). They have evolved from single measures of things like population to indexes (like the UN Human Development Index), and from activities to outputs and then outcomes (such as QALYs—quality adjusted life years—and DALYs—disability adjusted life years in relation to health). In all these ways, both states and societies watch themselves and recognize well-calibrated observation as the precondition for thought.

Every act of standardization makes thought and action on a large scale easier. But it also involves a loss of context and meaning, just as if the whole world spoke only one language, we'd miss out on the subtly different ways in which languages think. Every *ontology*—the term now used for information organizing tools—is selective, and large-scale databases are particularly selective.[2] Some methods, though, make it possible to combine the benefits of large-scale data sets, which can be easily manipulated, with sensitivity to widely varying local situations.[3] Opening data up, including raw data, collecting perceptions as well as objective facts, and encouraging communities of users to offer feedback to those doing the structuring can all mitigate the risks of overstandardization. So, for example, standardized repositories of "what works" can continuously encourage their users to ask the supplementary question, What works where, when, for whom, and why?[4] Standardized statistics can be questioned to see if they really capture what matters, so as to avoid the trap of only managing what's measured. In this way the standards can loop back on themselves, thinking intelligently about their own intelligence.

INTELLIGENT ARTIFACTS

Things are the second set of infrastructures essential for collective intelligence. The *OED* wasn't just an abstract set of classifications; it was embodied in pieces of paper. Better artifacts help any large-scale coordination and collaboration. We think with things and struggle to think without them. It's often easier to solve a problem if you can visualize it with a pen and paper, or a data visualization. Writing allowed classifications and abstractions, and then maps, charts, or tables to make the chaos of the world graspable and stable. The ruler helps us measure things in comparable ways. The consistent size of plates, forms of cutlery, or shapes of clothing makes everyday life easier. Cars that fit onto roads certainly help with mobility. Large-scale cooperation breeds large-scale standardization, and vice versa. Bar codes, URLs, and credit cards are just a few of the common things that assist the world to get along and think. Simplified digital interfaces make it easier to collaborate, whether to write software code for GitHub or organize a meeting by Facebook.

Even before the ubiquity of digital technologies, it was evident that thought could happen outside as well as inside the brain of the thinker. In some senses, thought is continuous with its environment. Think, for example, of a rider and the horse, or someone taking part in a particularly electric public meeting. In other senses, thought is closely coupled with things, like the books in a library or the signs that guide drivers around a city. The history of culture has constantly reshaped environments to help us think—with everything from weather vanes to door numbers—and in everyday life we rely on others—spouses and colleagues—to help us remember and think. Everyday intelligence is becoming ever more tightly coupled with the social and physical environment.[5] Prosthetic devices are penetrating bodies for sight, mobility, and perhaps soon memory (I have a near field communication implant in my hand that isn't yet useful, but points to a future where we all become partly cyborg), and sensors and calculating tools are animating the world around us. The idea that thought only takes place within autonomous, bounded individuals will appear ever more far-fetched.

Driving is a good illustration. Traffic deaths have declined sharply over the last ninety years. In the United States, for instance, still one of the more dangerous countries for drivers, fatalities per million miles driven fell from 24 in 1921 to 1.1 in 2015. They fell because of the combination of intelligence embodied in things (safer cars with better brakes and stronger materials), intelligence embodied in rules (road markings and speed limits), and intelligence embodied in people (who have to learn to drive). But the greatest advances in the next few decades will almost certainly come from embedding more intelligence in things, from driverless cars linked into smarter urban infrastructures.

INVESTMENT OF SCARCE RESOURCES IN INFRASTRUCTURAL CAPACITIES

Institutions that can concentrate time and resources over long periods are the third set of infrastructures for collective intelligence. Intelligence requires work, time, and energy—all of which could be devoted to other things. So collective intelligence depends on organizations as well as whole societies that are willing to devote scarce resources to building up human, social, and organizational capital, the knowledge capital held in libraries and professions, and the physical capital of intelligence embedded in machines. How well advanced these are determines what a society can do— how well it can thrive or survive.

Much of the crucial progress of the last few centuries resulted from improvements in all of these. In the late fifteenth century, less than 10 percent of the population in most of western Europe was literate. Over the next few centuries, however, literacy rates soared to 50 percent in some countries and have continued to rise with ever more years of education. The ability of societies to cooperate, trust, and share has also risen by all available measures, with crime falling by a factor of five between the fifteenth and eighteenth centuries.

The jump in capacity can be measured in the scale of institutions designed to think. There were large state bureaucracies at many points in history, from ancient Sumeria to Persia, Rome and China. But these only became common across the world during the last century, following a model pioneered by Europe. For example, the UK government's share of national income rose from around 1 to 2 percent in Elizabeth I's time in the sixteenth century to 10 to 20 percent two centuries later as state institutions gathered shape and used enormous levels of national debt to expand their reach. Later came the growth of the firm, charity, and university—all marshaling the brainpower of people on much larger scales. In the nineteenth century, the average factory grew from a typical size of under twenty workers at the beginning of the century to several hundred at the end, followed shortly after by an even greater growth in the size of businesses.

More complex and interdependent societies required more deliberate coordination. The early railways and roads brought with them disastrous crashes, prompting a search for better tools for control, which in turn required more investment and more concentration, manifest in the birth of the multidivisional corporation, large government agency, and regulator. Some societies consciously set themselves to work to increase their collective intelligence, buying the best of the world's brainpower and mobilizing their own, such as Germany in the nineteenth century under the influence of the economist Friedrich List, the United States after the Second World War under the influence of Vannevar Bush, and Singapore after its independence, through compulsory education, new institutions of research, and often-huge commitments of money.

It's now a commonplace that societies need to invest substantially in the skills of their people if they wish to thrive. But it's only quite recently that we have known more about the lost potential that comes from failing to do this adequately. An important research study demonstrated this in relation to inventive capacity. Analyzing 1.2 million inventors in the United States over a twenty-year period, the study showed the strong correlation between parental influence, direct experience of invention and quality of schooling, and the likelihood of a child ending up as an inventor. Children born into wealthy families where the parents worked in science and technology were far more likely to become inventors themselves. The implication was that the majority of bright children's potential was going to waste. This led the researchers to argue that shifting public resources from tax incentives toward subsidizing more opportunities for invention in early childhood would amplify the creative and economic potential of the United States.[6] The study also showed geographic effects: growing up in an area with large, innovative industries made it more likely that children would become inventors.

If collective intelligence depends on connecting the elements, just as individual intelligence depends on linking up neurons, it's not surprising that population density plays a part. It was no coincidence that the *OED* was born in the largest city in the world at the time. Cities have always been great crucibles of collective intelligence, mobilizing intensive interaction, through conscious design and the random serendipity of coffeehouses and clubs, societies and laboratories. Today it's estimated that doubling city population correlates on average with a 10 to 20 percent increase in wealth creation and innovation.[7] The reasons have to do with serendipitous interaction, but also with the growth of institutions that can mobilize the resources needed to broker and synthesize. These make the most of proximity, which matters to large-scale thought because we rely greatly on the subtle clues that come from face-to-face interaction. This is also why meetings persist in environments rich in digital technologies —the more disembodied communication is, the higher the ratio of miscommunication. So e-mails are more prone to misunderstanding than phone calls, which are in turn more prone to misunderstanding than direct conversation.

Societies of Mind

Networks are the fourth infrastructure for collective intelligence. Simon Schaffer brilliantly described Newton's *Principia Mathematica* as an example of what appears to be the product of a single mind, but is instead the product of a network of contacts, providing information and ideas on a large scale. Newton's thought happened in his brain, but also in his networks. The theory could not exist without the ties. Newton's correspondents were spread all over the Western world.

The modern world has depended on such networks—horizontal societies of mind that guide and nurture knowledge commons. The Royal Society was formed in London in 1660, and by 1880, there were 118 learned societies devoted to science and technology with nearly fifty thousand members, part of a "luxuriant proliferation of societies, associations, clubs, institutes and institutions."[8] Some were organized around general interest, and others around specialist subjects, hobbies, professional certifications, or advancing the frontiers of knowledge using journals, conferences, or mutual commentary to advance physics, design new ships, or improve the state of medicine.

The science system was similar, but on a much grander scale: peer-reviewed journal articles contributed to a truly cumulative knowledge—an idea based on simple design principles, yet with limitless scope to amplify and extend, capturing, ordering, and spreading what's known, combining wide networks with rigorous hierarchies to judge what to include and take notice of. The term *scientist* was only coined in 1833 (by William Whewell, who also incidentally coined the words *physicist, anode, cathode, consilience,* and *catastrophism*). But science then quickly made the transition from a field of gifted amateurs helped by patronage to become systematized, vast in scale, and highly collective in nature.

Scientists' ethos was and is different from that of the traders who prospered in parallel with them. The sociologist Robert Merton wrote of the "communism" of scientists: the belief that knowledge should be shared, not owned. Many papers in *Nature* now have more than 100 coauthors (and one recent physics paper had over 5,000), and there is an average of 8.4 coauthors on *Proceedings of the National Academy of Sciences* papers (double what it was in the 1990s).[9] Despite living in an individualistic culture, teams tend to dominate not just in science but also in business, technology development, and government, which is why the craft skills of how to recruit, motivate, and manage teams have become as important to the practice of collective intelligence as the hardware and software.[10]

The broader point is that science is inescapably collective. What counts as good or true is a social fact as much as a material one. Karl Popper put the point well: "Ironically enough, objectivity is closely bound up with the social aspect of scientific method, with the fact that science and scientific objectivity do not (and cannot) result from the attempts of an individual scientist to be 'objective' but from the friendly-hostile cooperation of many."[11] Indeed, the system can only function if the idea of private property and privacy is suspended—if new discoveries are revealed as a condition for the time-limited property rights conferred by patents or participation in the scientific community (an ethos that can be extended much more widely to the belief that the collective must always benefit from breaking down barriers of protection and privacy, the idea captured by Dave Eggers in his novel *The Circle* with the slogan "all that happens must be known").

The networks described above—like the Royal Society—existed in relation to hierarchies, including the navy and government, and ever-larger corporations and boards allocating funding. Here we see a general pattern. Widely distributed networks are good for argument and deliberation as well as gathering information. They feed and organize knowledge commons. But they work less well for decisions and integrating them for action. These roles tend to be dominated by small groups, with a strong mutual understanding. The *OED* had a similar pattern, as do more recent projects like Linux software or Wikipedia. They combine wide networks with much smaller core groups of guardians, curators, and editors who guide contentious decisions.

GRAND PROJECTS

The *OED* was a grand project, an assembly that combined observation, analysis, and memory. Just occasionally, societies have attempted even grander schemes to concentrate and amplify intelligence through explicitly organized infrastructures. The Manhattan Project to design a nuclear bomb during the Second World War was the largest such venture at the time. It employed seventy-five thousand people at its main site at Oak Ridge, Tennessee, and another forty-five thousand at Hanford, Washington, including many of the best physicists in the world, working in highly compartmentalized teams that often didn't know how their work fit into the bigger picture. Like other such ventures, it was also a model of collaboration—bringing together many different disciplines and coordinating networks of commercial contractors in the ultimately successful mission of building a workable atomic bomb.

NASA's work to land a person on the moon was even bigger, with well over four hundred thousand people employed on the Apollo program, and twenty thousand firms and universities involved as partners. These set the tone for the battle to conquer cancer and later map the human genome. Each had a specific goal, with the first two prompted by war and geopolitical competition. They benefited from their clear focus—a task with a beginning, middle, and hopefully end.

There have also been a few, uneven attempts to design whole systems that have in-built capacities for intelligence and were intended to last indefinitely. Cybersyn in Chile in the early 1970s is perhaps the best known. It was an example of vision and madness combined, a cybernetic system to plan an entire economy, devised several decades ahead of the many technologies that were needed for it to work. Under the auspices of the socialist president Salvador Allende, it was conceived as a distributed decision support system to help run the Chilean economy, with computer models to simulate how the economy might move, software to monitor factory production, a central operations room (with futuristic chairs similar in design to those used in *Star Trek*), and a network of telex machines all linked to a single mainframe computer. The idea was to help the economy think, as a system, but also to allow for more decentralization among factory workers, who could use the feedback of the system to manage themselves.

A US-backed military coup in 1973 cut this experiment short. But its animating ideas remain very much alive. The ideal of a system that can provide rich data and feedback to all participants so that they can better manage their own choices without dependence on an oppressive hierarchy can be found in many contemporary visions of the future of collective intelligence. Forty years later, cheap processing power and connectivity make this a much more plausible idea, and within many of the largest firms—like Amazon or Walmart—some of the Cybersyn ideas (stripped of the political ideals) are everyday realities.

Although many of the twentieth century's grand projects are inspiring, disturbing, and impressive in their different ways, perhaps the oddest feature that they share is how little influence they had, not how much. They dazzled, but then fizzled out. There was no second Manhattan Project. NASA shrunk. The mapping of the genome continued to offer great promise, but always some distance into the future. Each was a conscious attempt to orchestrate brainpower on a far larger scale than ever before. Yet they left the dominant, large-scale systems for collective intelligence relatively unchanged; how we organize democracy, governments, large firms, and universities has scarcely been influenced by the big projects. Instead, all these institutions would be easily recognizable to a visitor from a century ago.

- 5 -

The Organizing Principles of Collective Intelligence

WE HAVE SEEN THAT COLLECTIVE INTELLIGENCE depends on functional capabilities—like the ability to observe well—and is then supported by infrastructures, such as common rules. But how is it best organized? To answer this question, I start with three different examples of collective intelligence in the real world before turning to general principles that can be applied to any organization or group, and help explain why some so often act foolishly despite being full of clever people and machines.

THE ABANDONED PASSENGERS AT AN AIRPORT

A hundred passengers are stranded in a remote airport on an island. All the staff members disappear without explanation. The energy switches off, as do the mobile networks.

How do the abandoned passengers turn themselves from a collection of strangers into a collective intelligence, from a crowd into a group? Although this is an extreme illustration, we encounter versions of it all the time when people are thrown together, and have to work out ways to think and act as one, whether in an office, political party, or a neighborhood, or on a sports team.

The first task that the abandoned travelers face is how to communicate. If they share no spoken languages, they will have to improvise. Otherwise those sharing a language, and even better, those who are multilingual will quickly become powerful within the group. Then they will need to define a shared purpose. Is it to communicate with sources of help, perhaps on the mainland, or at another airport? Or it is to find ways to help themselves? Presumably both will matter.

Soon they will need facts—observations to make sense of their situation and prospects. Some of these facts will be close at hand. Is there enough food around, and how long will it last? Is a hurricane descending on them that might at least partly explain their predicament? Are they under an immediate threat, perhaps from terrorists or freezing temperatures? Other relevant information will be found in their collective memory. Are there lessons to be learned from similar incidents?

Then they need some creativity to generate options, such as how to find food, light, warmth, or help. And finally, they will need judgment, presumably through open conversation that interrogates and improves on the options that arise.

That may be hard. A few of them may set off on their own, convinced that the group is deluded or doomed. Others may just wait and pray. And the group may be riven by irreconcilable differences.

But this stylized example is a recognizable simulation of almost any joint human activity, here accelerating in a few hours what usually takes years or centuries. It's not dissimilar to what happened when thirty-three miners were trapped underground for two months in the San Jose mine in Chile (though they already knew each other and shared a language), when the Uruguayan football team survived for ten weeks in a remote area of the Chilean Andes after a plane crash, or in a different way, when two companies merge.

Our abandoned travelers will be greatly helped, as we are, by what lies around. Useful intelligence will be embodied in objects around them: battery-operated lights, food in tin cans, and perhaps, if they're really lucky, a battery-operated satellite phone. Intelligence will also be embodied in the people—their experience (Did any work in an airport?) or formal knowledge (Perhaps they will have to fly a plane in search of help?). And they will do much better if they are already good at trust—able to trust strangers and generate trust in each other. That will help them to devise some simple rules about who to listen to, who to care for first, and how to make decisions. These may help them to generate a rough-and-ready hierarchy, such as to make decisions about allocating scarce food supplies in a fair way.

What these add up to is a form of informational commons—a shared body of information and knowledge that the group owns, contributes to, and uses together. Without this they will be bereft. As a rough generalization, the more they can share knowledge and together verify which information to rely on, the better their chances of survival. That's why they'll need to talk, a lot (an interesting finding from research on situations like this is that nonstop talk, both vocal and nonverbal, is a crucial source of coordination in complex systems that are susceptible to catastrophic disasters).[1]

THE WORLD GRAPPLING WITH CLIMATE CHANGE

Our next thought experiment goes big. Imagine the whole of humanity could think as one. Imagine if our observations, thoughts, and feelings could be shared without any distortion, and if we could think together about our great challenges—from hunger to climate change—and work out solutions as if we were a single brain, helped by the vast array of technologies at our disposal.

This idea is a pure fantasy. Yet it's much closer to being real than a century ago, let alone five centuries ago, thanks to social media and the Internet. It's an ideal that in smaller form has resonated through history. Many groups tried to start from scratch, and leave behind the corruption and failure of the societies from which they came to become something closer to a true collective intelligence. The sects by the Dead Sea, the Buddhist and later Christian monasteries, the Pilgrims traveling to the United States, and the communes, cooperatives, and garden cities of the nineteenth century to the 1960s are all examples.

There are similar ideals of perfection within social science.[2] The economist Leon Walras suggested the idea of a perfect equilibrium in which no one could make any choices that would leave them happier. Perfect equilibrium rests on perfect competition and perfect information. Everyone's wants are expressed and then refracted through the market, which connects them to the economy's productive potential. Money is the currency for all desires, from the mundane to the exotic, and makes them all commensurable, manageable, and tangible. It's a vision compelling in its simplicity (however complex the mathematics), and seems to go with the grain of human nature and offer the prospect of an automatic mechanism endowed with perpetual motion.

The German philosopher Jürgen Habermas proposed a parallel ideal of perfect communication, again as a useful tool for interrogating imperfect reality. Our ideal of collective intelligence, or the united minds of the world, would have something like perfect information—accurate knowledge of its facts and circumstances along with perfect communication as well as the ability to share information, views, hopes, and fears with its members. It would also depend on trust, since against a background of hate or fear, no amount of information would make people cooperate.

This picture is abstract. But all of us will have experienced situations not so far from this ideal. Friends who are open, discuss together what to do, share tasks, and take turns experience a rough approximation of perfect community. With people we know well, we can communicate not just through our words but also through what we don't say. Perhaps it is this we aspire to in love as well. It is also not so far from what can be found in the happiest and healthiest families—taking turns, and sharing the good and bad helped by mutual empathy.

What methods of decision making and heuristics would a world trying to think as one use? It would need to be able to cope with multiple cognitive styles, from stories to facts, images to prose, and have many abilities (of analysis, observation, and judgment). It would probably gravitate neither to equality of voice nor fixed hierarchy, but rather to contingent inequality—giving greater voice to those with the greatest reputation, or who are the most admired and reliable. We can already see this in the world of the web: the strength of your voice depends on how many others want to listen to you. They may want to listen to you because of your authority or learning, but this isn't guaranteed. We might also expect the perfect community to recognize strength of feeling: how much you care about something affects how others respond to your hopes or concerns.

Seen in this light, our existing decision-making systems, like representative democracy or the market, turn out to be limited and special cases of collective decision making. The market uses binary decisions (whether or not to buy) and a single currency, money, while democracy uses the currency of votes between a handful of choices every few years. Yet we recognize that higher-bandwidth conversations are preferable, and more likely to help us achieve our goals.

A host of recent innovations are attempts to move in this direction, such as crowdsourcing, deliberative democracy, and open innovation. They point us, tentatively, toward a different future, where minds are partly integrated into a commons. It's an unsettling world for people brought up on the idea of the sovereign individual, masters of their own thoughts. But it also offers the promise of an intelligence on the same scale as the problems our world faces.

This abstract example becomes regularly real when the world thinks about truly big issues. Perhaps the biggest of them all, and one that is bringing interesting innovations in collective intelligence, is climate change. Their most visible forms are the gatherings of world leaders—in 2010 in Copenhagen, and then again in Paris in 2015—to agree on new treaties for climate change. The conferences had to distill the views of over two hundred countries, aligning politics, economics, and ecology. They used many of the techniques developed by diplomacy—unsuccessfully in 2010, and more successfully in 2015—including a huge amount of talk and preparation. But they also drew on one of the greatest exercises in orchestrated intelligence in human history: the Intergovernmental Panel on Climate Change (IPCC). The IPCC is an extraordinary assemblage of intelligence of different kinds, pooling data, using sophisticated supercomputers to model the weather, designing detailed scenarios, and mobilizing thousands of scientists to comment and critique. Around it are parallel attempts—like the C40 group of cities, Copenhagen Consensus conferences (which gather experts to rate the effectiveness of different solutions), Climate CoLab (which runs contests to find solutions), corporate attempts to create standard measurements of carbon emissions, and around all of them a swirl of NGOs, commentators, and expert groups that together constitute a global knowledge commons that turns knowledge about a shared threat, climate change, into a shared good.[3]

It's too soon to judge its success.[4] The IPCC appeared to get many things right: an autonomous system of orchestrating intelligence that couldn't be overly influenced by governments or big firms; some balance between its elements of observation, analysis, and memory; and a distribution of tasks among thousands of specialists.[5] But it could do relatively little to influence the authorizing environment around it and political pressures that influenced the main holders of power.

The Garage in a Small Town

Our third thought experiment is much smaller in scale. You can see it in a live form a few minutes from where you're sitting now. This is the challenge of sustaining a small garage—a group of mechanics working to fix cars in a small town or on the outskirts of a big city. Their intelligence is embodied in machines, learned through long apprenticeship that gives the mechanics a mix of formal knowledge and the ability to quickly assess problems. It's supported in manuals and guides. The environment they work in is in some respects quite stable—cars have changed surprisingly little in a century, and still have internal combustion engines, rubber, tires and padded seats. Most use gasoline to power them and oil to lubricate them. The mechanics' skill is a practical intelligence that's tested by whether the broken cars that come into their garage leave in a fit state to drive.

Yet in some respects, the environment faced by the garage changes all the time. The digital content of cars grows year by year; internal combustion engines are conceding ground to electrics and hybrids. New business models are now on offer, such as being able to lease cars or tools. And so like any organization, the mechanics running the garage need to choose what proportion of their time and energy to devote to the different dimensions of intelligence. How much time should they spend observing and scanning, how much energy should they devote to memory and files, and how much effort should they put into devising new offers? Just as important, they have to decide who to share with, how much of this knowledge to keep within the organization, and how much to pay for new knowledge when it's needed.

Just like the bigger examples cited before, the garage creates its own kind of knowledge commons: a body of knowledge about how things work that new staff members are inducted into, and whose quality determines whether the garages thrives or fails.

The Five Fundamental Principles for Organizing Collective Intelligence

What links these different thought experiments, each of which is also an everyday reality? What is it, at the micro and macro levels, that allows collective intelligence to flower?

Five crucial but nonobvious factors make all the difference. These factors may sound abstract. But they quickly become practical, giving shape to the knowledge commons that holds the group together, and they are relevant to any large-scale group wanting to think, act, and learn coherently and successfully. These are the *organizing principles* for collective intelligence.

The first is the extent of what I call the *autonomous commons* of the intelligence in the system. By this I mean how much the elements of intelligence are allowed free rein, and not subordinated too easily to ego, hierarchy, assumption, or ownership. Autonomy means allowing arguments to grow and become more refined. It requires a dialectical approach to intelligence—seeking out alternatives and refutations as a way of sharpening understanding. A group where people quickly become attached to their assertions, where secrets are guarded, or where too much weight is put on the speaker instead of what they say will tend to be collectively less intelligent. So will one that narrows options too quickly.

The second factor will be a contextually proportionate *balance*: how balanced the intelligence is between its different elements, and how well suited the balance is to the tasks at hand. Intelligence combines many distinct elements, from observation and focus to memory and creativity (described in chapter 3). Groups, like individuals, need to keep these in balance, and a high proportion of the cases where collective intelligence goes wrong reflect problems of imbalance, such as where groups are rich in data but poor in judgment, or rich in memory and poor in creativity, and vice versa. Knowing how to orchestrate these different elements of intelligence in a coherent way is one of the fundamental tasks facing any group and leader.

The third factor will be how well the group can *focus*. Focus means attending to what really matters and not being distracted. Knowing what to ignore matters as much as knowing what to attend to. That may not be so obvious. For the group stranded at an airport, there will certainly need to be a focus on getting through to someone out there. But if they are to be stuck for a long period, then holding the group intact and preventing conflict may matter even more. Focus also has a subtler meaning since it introduces granularity—knowing what is relevant on different scales.

The fourth factor will be the group's capacity to be *reflexive*—to be intelligent about itself and recursive.[6] Knowledge needs knowledge about the knowledge, and this requires loops—what I describe (in chapter 6) as the three loops of active intelligence: thinking about things, changing the categories with which we think about things, and changing how we think. The more reflexive any group is, the more intelligent it is in the long run. As I will show, this reflexiveness works best when it is most visible—for example, with predictions made explicitly, and explicit learning when

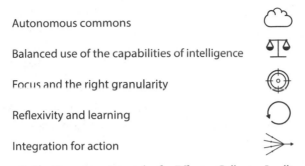

Autonomous commons

Balanced use of the capabilities of intelligence

Focus and the right granularity

Reflexivity and learning

Integration for action

Figure 3. The Organizing Principles for Effective Collective Intelligence

the anticipated doesn't happen, all feeding into a shared knowledge commons.[7] And it works best when it is helped by what I call *self-suspicion*—the ability to question the patterns that make most sense.

Finally, the fifth factor will be the group's ability to *integrate for action*, drawing on different types of data and ways of thinking to make a decision. It's not enough to think great thoughts and host glorious arguments. Life depends on action. So this type of integrative thinking is what marks out the most sophisticated civilizations. It's much of what we and our ancestors call wisdom, and it tends to develop through experience rather than only logic. It's where thought and action come together. We complicate to understand, but simplify to act, and search for a simplicity that lies on the far side of complexity.[8] What I call *high-dimensional choices*—complex in terms of cognitive tools, social relationships, and time—require more work and loops to arrive at a composite picture that can guide action, whether that action is physical in nature or communicational.

Together, these five organizing principles help any group to think more clearly about the past (the relevant collective memories), present (the facts of what is happening), and future (the options for resolving the situation). They help the group to imagine possible future options, discover them, and then realize them.

These dimensions of intelligence sound simple. But they are remarkably difficult to sustain. This is why intelligence is fragile and rare, and runs as much against nature as with it. Collective intelligence can easily regress—as has happened time and again in many places throughout human history, from Tasmanian aborigines to the once-great cities of Mohenjo Daro or Machu Picchu. The world is full of places where people live amid the ruins of superior past civilizations and full of institutions that were once far more competent than they are today. The direction of travel in our world is toward more complexity, more integration, and more intelligence—yet this is by no means a given.

Powerful enemies threaten intelligence all the time, such as purveyors of lies, distortions, rumors, and distractions, and in the digital environment, threats like trolling, spamming, cyberattacks, and denials of service. All can disrupt clear communication and thought.

The virtues that underpin collective intelligence are also rare and difficult, because each one clashes with other basic features of human interaction and other virtues. The autonomy of intelligence challenges the social order (which rests on an agreement not to see certain things while suppressing the views and voices of the powerless). It also runs directly against accountability in many instances (and as I'll show, too much accountability, like no accountability, can make institutions surprisingly stupid).

Balance challenges the status of the groups or professions associated with particular elements of intelligence, such as the guardians of memory. Reflexiveness challenges practicality and the pressure of events—the need to act now. It takes time to think, and time is scarce, so life rewards shortcuts.

Focus fights against curiosity, and it is particularly hard for clever people (and intelligent groups and civilizations have repeatedly been defeated by ones that are less intelligent but more focused on what really matters in a particular situation). Appropriate focus is even harder for machines that, like human brains, struggle to concentrate in a granular way, recognizing the scale of different tasks and contexts.

Finally, integrative thinking fights against our tendency to latch on to one way of thinking (to the person with a hammer, every problem looks like a nail).

This framework helps to explain why some fields act foolishly. The most common pattern is a failure to sustain the autonomy of intelligence. This can happen because power subordinates truth, but it can also happen because of incentives (as happens often in finance) or excessive loyalty.

The five principles for organizing collective intelligence are rarely found together in an ideal form. Yet most human groups follow some of them, however imperfectly. They provide a theoretical basis for understanding any kind of intelligence that exists at the level of a group, organization, network, or family. They mimic some of the properties of natural evolution and development. They involve the multiplication of options, selection, and replication (and like DNA, rarely mutate in wholly random ways, but tend to mutate more where there have already been mutations, or where there is stress and pressure). Nevertheless, the patterns of intelligence are also unlike the natural world, primarily because of awareness and freedom; we can choose whether to give greater weight to collective intelligence and can turn this into a moral choice—a commitment to be part of the intelligence of larger wholes.

- 6 -

Learning Loops

I EARLIER DESCRIBED THE ETYMOLOGIES of the words intelligence and collective, and showed that they have at their core a notion of choice within contexts of possibility and uncertainty. Any being faces an infinity of choices and no certainty about the future. We use all the elements of intelligence to help us understand what our choices really are, drawing on the limited data available to us as well as the mental models we have developed or acquired.

This mental task can be thought of in probabilistic terms. At every step, we try to make sense of the probability distribution of different outcomes. Are we at risk of attack? Will it rain? Will my friend still be my friend? If I build my home in this way, will it survive a storm? To make sense of these choices, any intelligence has to assess the possibility space lying ahead of it and the likely probability distribution. Is the context similar to one we've encountered before? Do our existing categories or concepts apply?

Our default is to depend on what we already know and change only incrementally, in small adjacent steps. This is the logic of evolution, in which big changes tend to be the consequence of many small changes rather than giant leaps.

The same is true of learning. Some learning is algorithmic, some is experimental, and much is sequential—what you can learn depends on what you have already learned. The computer scientist Lesley Valiant writes of such tools as "eliminations algorithms" and "Occam," and what he calls "ecorithms" that help an organism cope with an environment by learning. "Cognitive concepts," he writes, "are computational in that they have to be acquired by some kind of algorithmic learning process, before or after birth. Cognitive concepts are, equally, statistical in that the learning process draws its basic validity from statistical evidence—the more evidence we see for something the more confident we will be in it." These models and inputs are considered in what has variously been called the mind's eye, ego tunnel, or conscious present, where new data intersect with our longer-term memories, the moment between a known past and unknown future.

The mark of any intelligent creature, institution, or system is that it is able to learn. It may make mistakes, but it won't generally repeat them. That requires an ability to organize intelligence into a series of loops, which have a logical and hierarchical relationship to each other.

First-loop learning is what we recognize as everyday thought. It involves the application of thinking methods to definable questions, as we try to analyze, deconstruct, calculate, and process using heuristics or frameworks. We begin with models of how the world works as well as models of thinking, and then we gather data about the external and internal worlds, based on categories. Then we act and observe when the world does or does not respond as expected, and adjust our actions and the details of our models in response to the data.

These first-loop processes of interpretation and action are imperfect. Much is known about confirmation biases along with our failure to think probabilistically or logically. But the first loop helps to correct our intuitions (what Daniel Kahneman calls System 2 processes of considered thought, helping to correct the otherwise-everyday use of System 1 intuitions). This kind of thinking helps us get by most of the time. It is functional, practical, useful, and relatively easy. The combination of facts and models is what enables life to function, and is how our brains work most of the time.

Within organizations, explicit processes for learning can dramatically improve performance. Later, I will discuss the procedures used in the airline industry to learn from crashes or near misses, hospitals that regularly review data and lessons learned, and factories that empower workers to fix problems. What's more remarkable is how many institutions lack even basic learning loops of this kind, and so continue to make unnecessary mistakes, assume facts that aren't true, and deny the obvious.

Second-loop learning becomes relevant when the models no longer work or there are too many surprises. It may be necessary to generate new categories because the old ones don't work (imagine a group that has moved from a desert environment to a temperate mountain zone), and it may be necessary to generate a new model, for example to understand how the stars move. This second loop also involves the ability to reflect on goals and means.

This is often what we mean by creativity: seeing in new ways, spotting patterns, and generating frames. Arthur Schopenhauer wrote that "the task is not so much to see what no one yet has seen but to think what nobody yet has thought about that which everybody sees." Saul Bellow was implying something similar when he spoke of the role of art as something that alone can penetrate some of the "seeming realities of this world. There is another reality, the genuine one, which we lose sight of. This other reality is always sending us hints, which without art, we can't receive. . . . Art has something to do with an arrest of attention in the midst of distraction." That arrest, a slowing down of thought and then a speeding up, takes us to a new way of categorizing and modeling the world around us. In handling high-dimensional problems, we frequently try to accumulate multiple frames and categories, see an issue from many angles, and then keep these all in mind simultaneously. This is hard work, and the hardness of the work rises exponentially with the number of frames in play simultaneously.

The relationship between first- and second-loop learning is fuzzy. Sometimes we have to take risks to find new ideas and new categories, even when our current models appear to be working well. This is the well-known trade-off between exploitation and exploration. Exploitation of what we already know is predictable and usually sensible. But if we never explore, we risk stagnation or at least missing out on new opportunities. So to thrive, we have to sometimes take risks, accept failures and bad decisions, and deliberately go off track and take a route that appears less optimal. Think, for example, of trying out a new restaurant rather than one we know and love. Because exploration is so essential to learning, a surprising finding of research on decision making is that people who are inconsistent sometimes end up performing better than ones who are consistent.

Third-loop learning involves the ability to reflect on and change how we think—our underlying ontologies, epistemologies, and types of logic. At its grandest, this may involve the creation of a system of science, or something like the growth of independent media or spread of predictive analytics.

We recognize third-loop learning to have happened when a radically new way of thinking has become normal, with its own tools and methods, and its own view of what is and what matters.

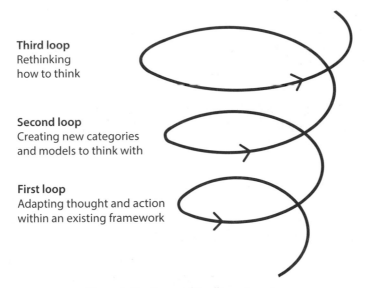

Third loop
Rethinking
how to think

Second loop
Creating new categories
and models to think with

First loop
Adapting thought and action
within an existing framework

Figure 4. The Loops of Intelligent Learning

Most fundamental social change also involves just such third-loop learning—not just doing new things to others. This is the implication of Audre Lorde's famous comment that we cannot use the master's tools to dismantle the master's house (and I examine it in more detail in chapter 16).

But we also see third-loop learning at a more mundane level, when individuals decide to live and think in a different way, such as by committing themselves to regular meditation.

All recognizably intelligent people and groups can adjust their behavior in response to surprises, adjust the categories they use more occasionally, and on rare occasions, adjust the ways in which they think.[1] Indeed, the psychological growth of any individual involves passing through all three loops repeatedly, as the individual learns about the world, and also reevaluates their place in the world and how to conceive of it.[2] Similar patterns are visible in groups and organizations, with the great majority of activity taking place within the first loop, the occasional use of second-loop learning to generate new categories and frames, and much more occasionally, a change to the whole cognitive model.

These basic characteristics of learning—iterative, driven by error and surprise, and with a logical flow from the small to the large—have been more fully embraced in some fields and societies than others. The acceptance of error in science along with the encouragement of surprise and discovery are fundamental traits of rational, enlightened societies. Being adept at all three loops helps the individual or organization to cope with multiple types of thought, choosing the right tool for the right task. Science advances through designing and testing hypotheses and theories. Philosophy involves questions—as Immanuel Kant put it, "Every answer given on principle of experience begets a fresh question." Art involves exploration.

Our brains work through analogy, metaphor, and the search for commonalities as well as through linear logic. For example, a good musical understanding can neither be acquired nor demonstrated by setting out arguments. It is instead best displayed and learned by playing music with feeling and understanding, and makes sense only in the context of a culture. Indeed, meaning only arises from cultures and large-scale uses rather than from simple correspondence of the kind that a search engine provides, and we can all recognize the difference between knowledge that is general (like the theories of physics) and knowledge that is by its nature particular, like the qualities of a particular person, poem, or tree.

Someone capable only of linear logic or first-order learning can appear dumb even if in other respects they are clever. So far, computing has proven better at these first-loop tasks than it has at second- or third-order learning. It can generate answers much more easily than it generates questions. Computers are powerful tools for playing chess, but not for designing games. Networked computers can help shoppers find the cheapest products or simulate a market, but offer little help to the designers of economic or business strategy. Similarly networked computing can help get people onto the streets—but not to run a revolution.[3]

This is territory where rapid advances *may* be possible; already we have many tools for generalizing from observed paired judgments as well as pattern recognition and generation, and huge sums are being invested in new forms of computing. For organizations, the challenge is to structure first-, second-, and third-order learning in practical ways, given scarce resources. The simple solution is to focus on first-order learning with periodic scans, helped by specialists, to review the need for second- or third-order learning, such as for changing categories or frames. We can visualize this as the combination of lines with loops—straight lines, or focused thinking combined with periodic loops to assess, judge, or benchmark.

This approach also helps to make sense of how all practical intelligence strikes a balance between inclusion and exclusion. Some types of data and patterns are attended to; others are ignored. This selectivity applies to everything from natural language to sensors. More is not always better; more inputs, more analysis, and even more sophistication can impede action. Practical intelligence has to select. There is a parallel issue for machine intelligence. Selection is vital for many complex processing tasks that are too big for even the fastest supercomputers if tackled comprehensively. This is why so much emphasis in artificial intelligence has been placed on selection heuristics or Bayesian "priors" that help to shrink the pool of possibilities to attend to.

Again, we complicate to understand and simplify to act. We may use wide networks for gathering information about options, and then use a small group or individual to make decisions in uncertainty.

Understanding better how focus is managed may turn out to be key to understanding collective intelligence—and how a shifting balance is achieved between wide peripheral vision and attention to many signals along with the focus needed for action.

But it's intriguing to reflect that there can be no optimum balance between these three loops. In principle, any intelligent group needs a capacity for all three. It's impossible, though, to know what the right balance between them should be. It's easy to imagine an organization locked only into first-loop learning (many banks or firms like Enron have been). Yet it's also possible to imagine organizations devoting too much scarce leadership time to second- and third-loop learning, reinventing their cognitive maps at a cost in terms of present performance.

In stable environments, the first-loop reasoners will tend to do best. In unstable ones, where any item of knowledge has a half-life of decay, the groups with more capacity to reimagine their categories and thinking modes may adapt better. In principle, any group should optimize for stable environments with a well-suited division of labor, until signs of change appear, at which point they should devote more to scanning, rethinking options and strategies, and mobilizing resources so that these can be redirected to opportunities and threats. But it is inherently impossible to know what the best balance is except in retrospect.

- 7 -

Cognitive Economics and Triggered Hierarchies

THE ORGANIZATION OF THOUGHT as a series of nested loops, each encompassing the others, is a general phenomenon. It can be found in the ways in which intelligence is organized in our bodies and within the groups we're part of. The efficient deployment of energy to intelligence depends on a similar logical hierarchy that takes us from the automatic and mindless (which require little energy) to the intensely mindful (which require a lot). With each step up from raw data, through information to knowledge, judgment, and wisdom, quantity is more integrated with qualitative judgment, intelligence becomes less routine and harder to automate, and crucially, the nature of thought becomes less universal and more context bound. With each step up the ladder, more energy and labor are required.

What Is the Organization in Self-Organization?

Seeing large-scale thought through this lens provides useful insights into the idea of self-organization. The popularity of this idea reflects the twentieth-century experience of the limits of centralized, hierarchical organizations—even if the world is still dominated by them, from Walmart and Google to the People's Liberation Army and Indian Railways. We know that a central intelligence simply can't know enough, or respond enough, to plan and manage large, complex systems.

Widely distributed networks offer an alternative. As with the Internet, each link or node can act autonomously, and each part of the network can be a fractal, self-similar on multiple scales.

There are obvious parallels in human systems. The term *stigmergy* has been coined to describe the ways in which communities—such as Wikipedia editors or open-software programmers—pass tasks around in the form of challenges until they find a volunteer, a clear example of a community organizing itself without the need for hierarchy.[1]

Friedrich Hayek gave eloquent descriptions of the virtues of self-organization, and counterposed the distributed wisdom of the network to the centralized and hierarchical wisdom of science or the state: "It is almost heresy to suggest that scientific knowledge is not the sum of all knowledge. But there is a body of very important but unorganized knowledge: the knowledge of the particular circumstances of time and place. Practically everyone has some advantage over all others because he possesses unique information of which beneficial use might be made, but of which use can be made only if the decisions depending on it are left to him or are made with his . . . cooperation." More recently, Frederick Laloux wrote the following lines, capturing a widely held conventional wisdom: "Life in all its evolutionary wisdom, manages ecosystems of unfathomable beauty, ever evolving towards wholeness, complexity and consciousness. Change in nature happens everywhere, all the time, in a self-organizing urge that comes from every cell and every organism, with no need for central command and control to give orders or pull levers." Here we find the twenty-first-century version of the late nineteenth-century notion of the élan vital, a mystical property to be found in all things.

It's an appealing view. But self-organization is not an altogether-coherent concept and has often turned out to be misleading as a guide to collective intelligence. It obscures the work involved in organization and in particular the hard work involved in high-dimensional choices. If you look in detail at any real example—from the family camping trip to the operation of the Internet, open-source software to everyday markets, these are only self-organizing if you look from far away. Look more closely and different patterns emerge. You quickly find some key shapers—like the designers of underlying protocols, or the people setting the rules for trading. There are certainly some patterns of emergence. Many ideas may be tried and tested before only a few successful ones survive and spread. To put it in the terms of network science, the most useful links survive and are reinforced; the less useful ones wither. The community decides collectively which ones are useful. Yet on closer inspection, there turn out to be concentrations of power and influence even in the most decentralized communities, and when there's a crisis, networks tend to create temporary hierarchies—or at least the successful ones do—to speed up decision making. As I will show in chapter 9, almost all lasting examples of social coordination combine some elements of hierarchy, solidarity, and individualism.

From a sufficient distance, almost anything can appear self-organizing, as variations blur into bigger patterns. But from close-up, what is apparent is the degree of labor, choice, and chance that determines the difference between success and failure. The self-organization in any network turns out to be more precisely a distribution of degrees of organization.[2]

COGNITIVE ECONOMICS

The more detailed study of apparently self-organizing groups points toward what could be called a *cognitive economics*: the view of thought as involving inputs and outputs, costs and trade-offs. This perspective is now familiar in the evolutionary analysis of the human brain that has studied how the advantages of an energy-hungry brain, which uses a quarter of all energy compared to a tenth in most other species, outweighed the costs (including the costs of a prolonged childhood, as children are born long before they're ready to survive on their own, partly an effect of their large head size).

Within a group or organization, similar economic considerations play their part. Too much thought, or too much of the wrong kind of thought, can be costly. A tribe that sits around dreaming up ever more elaborate myths may be easy pickings for a neighboring one more focused on making spears. A city made up only of monks and theologians will be too. A company transfixed by endless strategy reviews will be beaten in the marketplace by another business focused on making a better product.

Every thought means another thought is unthought. So we need to understand intelligence as bounded by constraints. Cognition, memory, and imagination depend on scarce resources. They can be grown through use and exercise, and amplified by technologies. But they are never limitless.

This is apparent in chaotic or impoverished lives, where people simply have little spare mental energy beyond what's needed for survival. As a result, they often make worse choices (with IQ falling by well over ten points during periods of intense stress—one of the less obvious costs of poverty). But all of us in daily life also have to decide how much effort to devote to different tasks—more for shopping or your job; more time finding the ideal spouse, career or holiday, frequently with options disappearing the longer you take.

So we benefit from some types of decision becoming automatic and energy free, and using what Kahneman called System 1 and 2. Walking, eating, and driving are examples that over time become automatic. With the passage of time, we pass many more skills from the difficult to the easy by internalizing them. We think without thinking—how and when to breathe, instinctive responses to danger, or actions learned in childhood like how to swim. We become more automatically good at playing a tune on the piano, kicking a football, or riding a bicycle. Learning is hard work, but once we've learned the skill, we can do these things without much thought. There are parallels for organizations that struggle to develop new norms and heuristics that then become almost automatic—or literally so when supported by algorithms. This is why so much effort is put into induction, training, and inculcating a standardized method.

Life feels manageable when there is a rough balance between cognitive capacity and cognitive tasks. We can cope if both grow in tandem. But if the tasks outgrow the capacity, we feel incapable. Similarly, we're in balance if the resources we devote to thinking are proportionate to the environment we're in. The brain takes energy that would otherwise be used for physical tasks like moving around. In some cases, evolution must have gone too far and produced highly intelligent people who were too weak to cope with the threats they faced. What counts as proportionate depends on the nature of the tasks and especially how much time is a constraint. Some kinds of thought require a lot of time, while others can be instantaneous. Flying aircraft, fighting battles and responding to attack, and flash trading with automated algorithmic responses are all examples of quick thought. They work because they have relatively few variables or dimensions, and some simple principles can govern responses.

Compare personal therapy to work out how to change your life, a multistakeholder strategy around a new mine being built in an area lived in by aboriginal nomads, or the creation of a new genre of music. All these require by their nature a lot of time; they are complex and multilayered. They call out for many options to be explored, before people can feel as well as logically determine which one should be chosen.

These are much more costly exercises in intelligence. But they happen because of their value and because the costs of not doing them are higher.

Here we see a more common pattern. The more *dimensional* any choice is, the more work is needed to think it through. If it is cognitively multi-dimensional, we may need many people and more disciplines to help us toward a viable solution. If it is socially dimensional, then there is no avoiding a good deal of talk, debate, and argument on the way to a solution that will be supported.[3] And if the choice involves long feedback loops, where results come long after actions have been taken, there is the hard labor of observing what actually happens and distilling conclusions. The more dimensional the choice in these senses, the greater the investment of time and cognitive energy needed to make successful decisions.

Again, it is possible to overshoot: to analyze a problem too much or from too many angles, bring too many people into the conversation, or wait too long for perfect data and feedback rather than relying on rough-and-ready quicker proxies. All organizations struggle to find a good enough balance between their allocation of cognitive resources and the pressures of the environment they're in. But the long-term trend of more complex societies is to require ever more mediation and intellectual labor of this kind.

This variety in types of intelligence, the costs they incur, and the value they generate (or preserve) gives some pointers to what a more developed cognitive economics might look like. It would have to go far beyond the simple frames of transaction costs, or traditional comparisons of hier-archies, markets, and networks. It would analyze the resources devoted to different components of intelligence and different ways of managing them—showing some of the trade-offs (for example, between algorithmic and human decision making) and how these might vary according to the environment.[4] More complex and fast-changing environments would tend to require more investment in cognition. It would also analyze how organizations change shape in moments of crisis—for instance, moving to more explicit hierarchy, with less time to consult or discuss, or investing more in creativity in response to a fast-changing environment.

Economics has made significant progress in understanding the costs of finding information, such as in Herbert Simon's theories of "satisficing," which describe how we seek enough information to make a good enough decision. But it has surprisingly thin theories for understanding the costs of thought. Decision making is treated largely as an informational activity, not a cognitive one (though greater attention to concepts such as "organization capital" is a move in the right direction).

A more developed cognitive economics would also have to map the ways in which intelligence is embodied in things—the design of objects, cars, and planes—and in systems—water, telecommunications, and transport systems—in ways that save us the trouble of having to think.

It would need to address some of the surprising patterns of collective intelligence in the present, too, many of which run directly counter to conventional wisdom. For example, organizations and individuals appear to be investing a higher, not lower, proportion of their wealth and income in the management of intelligence in all its forms, particularly those operating in competitive environments. Digital technologies disguise this effect because they have dramatically lowered the costs of processing and memory. But this rising proportion of spending appears close to an iron law, and may be a hallmark of more advanced societies and economies. Much of the spending helps to orchestrate the three dimensions of collective intelligence: the social (handling multiple relationships), cognitive (handling multiple types of information and knowledge), and temporal (tracking the links between actions and results).

A related tendency is toward a more complex division of labor to organize advanced forms of collective intelligence. More specialized roles are emerging around memory, observation, analysis, creativity, or judgment, some with new names like SEO management or data mining. Again, this effect has been disguised by trends that appear to make it easier for anyone to be a pioneer and for teenagers to succeed at creating hugely wealthy new companies. Linked to this is a continuing growth in the numbers of intermediaries helping to find meaning in data or link useful knowledge to potential users. This trend has been disguised by the much-vaunted trends toward disintermediation that have cut out a traditional group of middlepersons, from travel agencies to bookshops. But another near iron law of recent decades—the rising share in employment of intermediary roles, and the related rise of megacompanies based on intermediary platforms such as Amazon or Airbnb—shows no signs of stopping. In each case, there appear to be higher returns to investment in tools for intelligence.

A cognitive economics might also illuminate some of the debates under way in education, as education systems grapple with how to prepare young people for a world and labor market full of smart machines able to perform many more mundane jobs. Schools have not yet adopted

Jerome Bruner's argument that the primary role of education is to "prepare students for the unforeseeable future." Most prefer the transmission of knowledge—and in some cases rightly so, because many jobs do require deep pools of knowledge. But some education systems are concentrating more on generic abilities to learn, collaborate, and create alongside the transmission of knowledge, in part because the costs of acquiring these traits later on are much higher than the costs of accessing knowledge. The traits generally associated with innovation—high cognitive ability, high levels of task commitment, and high creativity—which were once thought to be the preserve of a small minority, may also be the ones needed in much higher proportions in groups seeking to be collectively intelligent.[5]

An even more ambitious goal for cognitive economics would be to unravel one of the paradoxes that strikes anyone looking at creativity and the advance of knowledge. On the one hand, all ideas, information, and thoughts can be seen as expressions of a collective culture that finds vehicles—people or places that are ready to provide fertile soil for thoughts to ripen. This is why such similar ideas or inventions flower in many places at the same time.[6] It is why, too, every genius who, seen from afar, appears wholly unique looks less exceptional when seen in the dense context of their time, surrounded by others with parallel ideas and methods. Viewed in this way, it is as odd to call the individual the sole author of their ideas as it is to credit the seed for the wonders of the flowers it produces. That some upbringings, places, and institutions make people far more creative and intelligent than others proves the absurdity of ascribing intelligence solely to genes or individual attributes.

But to stop there is also untenable. All thought requires work—a commitment of energy and time that might otherwise have been spent growing crops, raising children, or having a drink with friends. Anyone can choose whether to do that work or not, where to strike the balance between activity and inertia, engagement and indolence. So thought is always both collective and individual, both a manifestation of a wider network and something unique, both an emergent property of groups and a conscious choice by some individuals to devote their scarce time and resources. The interesting questions then center on how to understand the conditions for thought. How does any society or organization make it easier for individuals to be effective vehicles for thought, to reduce the costs and increase the benefits? Or to put it in noneconomic language, how can the collective sing through the individual, and vice versa?

The current state of understanding these dynamics is limited. We know something about clusters and milieus for innovation and thought. It's clearly possible for the creative and intellectual capability of a place to grow quickly, and using a combination of geography, sociology, and economics, it is easy to describe the transformation of, say, Silicon Valley, Estonia, or Taiwan. Yet there are few reliable hypotheses that can make predictions, and many of the claims made in this area—for example, about what causes creativity—have not stood up to rigorous analysis. For now, this is a field with many interesting claims but not much solid knowledge.

TRIGGERED HIERARCHIES AND CORRECTIVE LOOPS

Many groups—including apparently self-organizing ones—handle their tasks with a pattern that can be best described as "triggered hierarchy." Tasks are dealt with at a low level, with the maximum standardization and least thought or reflection, and thus least time and energy, until they don't work or bump into a problem. They are then dealt with at a higher level of the hierarchy or system, which requires more energy and usually more time.

Think, for instance, about how your body is regulated. Most of its processes are scarcely visible to you. Your body automatically maintains blood temperature and pressure. Then when you get sick your first response is to try to deal with it yourself—taking pills or going early to bed. Then if these don't work you go the doctor, and they in turn either deal with the symptoms or escalate to higher-tier, more specialized knowledge.

The patterns of everyday economic activity are similar. Much of it is effectively automated—people do their jobs in familiar ways, markets link supply and demand, and prices are set automatically. But when things go wrong, higher tiers of authority are brought in. This happens within individual firms, where incidents pull in higher levels of management. Some incidents will also pull in the state; an accident may bring in the local health and safety regulators, a fraud brings in the police, and an economic crash brings in emergency funds from the state. Seen as a system, the economy is constantly correcting, but it would be misleading to see this as self-correction. It is instead a combination of self-correction and correction by successively higher tiers of authority. The classic Hayekian account of a market self-organizing through price signals and other information flows is in this sense both ahistorical and misleading as theory: it's accurate for a proportion of the time, in normal circumstances, but fails to explain much of the system's intelligence, which manifests itself when surprises occur. This is what General Motors founder Alfred Sloan meant when he wrote in 1924 that his "senior team . . . do not do much routine work with details. They never get up to us. I work fairly hard, but on exceptions."

The account of corrective loops describes a general principle of effective human organization though it is significantly different from the classic accounts of law-based government or markets. It suggests that everyday processes are automatic and based on the accumulated intelligence of experience, but when trigger events happen—an illness, emergency, or crash—higher-tier authorities are pulled in, bringing with them additional resources, power, and knowledge. The same principles can be found in policing and criminal justice, the everyday life of a family, or the management of a manufacturing process.

Air safety is a particularly good example because it has strong systems for spotting errors and interpreting them and then generalizing the solutions. There can be many causes of a disaster—ice, lightning, terrorism, pilot suicide, or engine failure—each of which requires different preventive actions. The Aviation Safety Reporting System is a voluntary, confidential incident reporting system used to identify hazards that's run independently from the formal regulation of air travel (by the Federal Aviation Agency in the United States and equivalents elsewhere) and is therefore a good illustration of autonomous intelligence. Lots of data were collected before the Aviation Safety Reporting System, but they were kept hidden and so rendered largely useless, for fear of opening up risks of litigation. Other relevant methods include crew management policies that encourage junior pilots to challenge senior ones where risks are involved and through regular simulations of disasters that help staff to respond automatically when an emergency happens. Meanwhile, at the systemic level, initiatives like the European Strategic Safety Initiative try to implement improvements (say, on how to respond to volcanic ash).

The aim is to build multiple defenses against multiple threats, sometimes inspired by the "Swiss cheese theory" of risk, in which systems are likened to multiple slices of Swiss cheese, stacked on top of each other. Threats materialize when the holes are aligned, and conversely they are stopped when multiple defenses are layered on top of each other. So, for example, when on August 2, 2005, Air France flight 358 crashed while landing in Toronto, the crew members were able to evacuate over three hundred panicking passengers in less than two minutes, shortly before the plane burst into flames, with their training serving to mitigate the physical failure.

Safety at the level of the system thus depends on all three loops. A single aircraft crash leads to a tightening up of procedures and checklists (first loop). A spate of aircraft crashes with apparently similar causes may lead to a series of design changes to how airplanes are made, addressing some new category of error, such as cybersecurity or smarter terrorism (second loop). Stagnation in the aerospace industry or a failure to respond to a deeper pressure like the need to reduce carbon emissions may prompt a new model of thought, such as the invention of a new reporting system or use of open innovation methods to bring in ideas from new sources (third loop).

Organizational hierarchies often struggle to operationalize all three loops because the latter two are so likely to threaten the status of leaders or specialists. They raise questions, and require criticism and skepticism. Networks don't fare much better, lacking the resources to seriously rethink when things go wrong or to invest in alternatives. But the combination of hierarchies and networks, when helped by systematic looped learning, can be effective at making a system collectively intelligent.

Hospitals and health care systems have attempted to use some of the same methods, especially nonpunitive reporting of errors and adverse events as well as measuring and improving teamwork.[7] John Dewey described habit as the mediation between impulse and intelligence; these methods all try to turn shared intelligence into shared habits.

The famous Narayana Hrudayalaya hospital in Bangalore, for instance, gathers data systematically on operations and patterns, brings its doctors together for weekly meetings to discuss what has and hasn't worked, and then tries to implement and embed the lessons in tightly specified protocols for action. In a few cases, schools take a similar approach, with regular study circles to review new knowledge and take stock of achievements. But few systems have anything remotely as rigorous as air travel, which is why most are so much more collectively stupid than they could be.

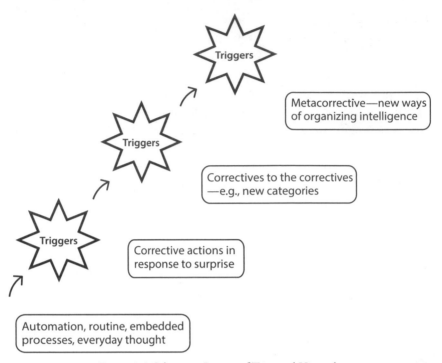

Triggers

Metacorrective—new ways of organizing intelligence

Triggers

Correctives to the correctives —e.g., new categories

Triggers

Corrective actions in response to surprise

Automation, routine, embedded processes, everyday thought

Figure 5. A Schematic Picture of Triggered Hierarchy

Within science there have been some attempts to formalize a parallel system of propositions, tests, and refutation. The work of Karl Popper and Imre Lakatos showed how science could move through all three loops— testing ideas within a given paradigm, then in some situations changing categories to generate new tests, and in a further set of situations redesigning the very framework for conceiving of scientific knowledge. The starting point is Popper's principle of refutability, but this is not enough to cope with new types of knowledge.[8]

Power can be defined as the ability to get away with mistakes.[9] These systems make it harder to get away with or hide errors, recognizing the collective interest in learning from them. They privilege intelligence over power, and the system's interest in learning over the individual institution's interest in keeping errors hidden. In many, if not most, real human systems, the hierarchy can be the enemy of intelligence, introducing intrusive intervention from higher-tier authorities that lack either the knowledge or motivation to serve the needs of the system. Yet the theory of triggered hierarchy points to how the different levels of a system can align themselves more constructively.

Drawing on these examples, we can describe a general model of intelligent governance in systems of triggered hierarchies. In this model there are multiple layers of governance on successively higher scales, responsible for the effective management of systems of ever-larger scale. Each successive layer has some defined authority, license, or power to intervene on lower ones, or hold them to account. These rights only become operational once agreed-on negative triggers happen. In more hierarchical systems, authority flows downward, from top to bottom, as in the classic empire or nation-state. But in the more democratic versions, each layer is accountable collectively to the lower ones over longer timescales than the timescale of individual decisions.

Patterns of failure and success generate adaptations and learning in both the intervention methods used and the patterns of triggering; in other words, second- and third-loop learning kick in. Invisible hands depend on visible ones; governance is both conscious and self-conscious, and can be made explicit. More complex, more interrelated, and higher-stakes problems require intensive cognitive labor—involving smaller, longer-lasting groups and higher levels of trust.

This theory of triggered hierarchies is an approximation. In many systems, there are few intermediate steps or the roles of the different tiers are ambiguous. The famous General Electric workout groups created by Jack Welch were designed to help people on the factory floor bypass the structured system of triggers—because too much vital feedback was being screened out. The higher levels may have less power but more influence, as in the air safety example. But in sophisticated, well-functioning systems, we almost always find something similar to the model described above, and most systems would work much better with more structured, systematic, and visible loops of this kind.

This may seem obvious. Yet this model of triggered hierarchies suggests that most of the common categories for thinking about decision making are inadequate. The classic array of options described in political science and economics polarizes the organizational choices in misleading ways that are too static to describe the real life of intelligent systems. On the one hand, there is the classic hierarchy, with authority cascading down through layers in the big corporation, government bureaucracy, or religious order. On the other hand, there is the invisible hand of the market. Believers in centralization contend with believers in decentralization.

If we observe successful human processes, though, we see that these are intermingled. No hierarchy can in practice specify and control every detail; instead, it tries to define boundaries of competence and authority for lower tiers, and lets them get on with things however much it may want to interfere and micromanage. Similarly, no real market is self-governing. All in practice coevolve with regulations, laws, and monetary authorities, which respond to triggers to maintain equilibrium. The most important question is the quality of those responsible for activating and responding to triggers: How good are they at recognizing a problem they cannot solve, and how open are they to learning even when that learning may threaten their status?

Intelligence, Representation, and Character

This approach to intelligence as structured into layers of links to the long-standing philosophical argument between the Cartesian view of thinking, as always defined by representation and linear logic (a view greatly extended by modern cognitive science), and an alternative view, expressed by Martin Heidegger among many others, that emphasizes intelligence that isn't manifest in this sense but rather becomes embodied.[10] For instance, when we use a tool like a hammer, we don't think through a representation but rather become one with the hammer. Our knowledge about the hammer reveals itself through the easy way in which we use it and is, strictly speaking, thoughtless. There's a parallel pattern in artificial intelligence and robotics, which can produce intelligent actions without representations, consciousness, or reasoning.[11]

But these types of intelligence, which are common, reinforce the points made earlier. We usually learn them through representation—copying someone else and repeating the movement until it becomes natural. At that point it is internalized and even embodied in us—so that we don't have to think about how we ride a bicycle, hit a hammer, or play the piano. It has become an automated routine, which we then only think about when something goes wrong—hitting a tree or a wrong note.

Embodiment is another word for character, the makeup of animals, institutions, or individuals. Some of that is rooted in genes and their interaction with environments. It gives us dispositions, the ways of seeing the world that make a tiger see things so differently from a mouse or an aristocrat understand the world so differently from a factory worker. It is shaped by early childhood, degrees of security, and reinforcement. Character also becomes important to this story since our habits become us—an ancient observation of philosophy. We do not so much choose directly, as choose indirectly through the habits we accumulate. We create ourselves through what we make automatic and instinctive, the accumulated results of experience that become aspects of character and intelligence.[12] That is why so often "we know more than we can tell," and can do such things as recognize a flying bird from the quickest of glimpses or break an egg in ways we could never satisfactorily explain.[13]

- 8 -

The Autonomy of Intelligence

Why very often, or nearly always, [are] the accidental images
the most real? Perhaps they've not been tampered with by the
conscious brain.
—*Francis Bacon, interview with David Sylvester*

IN THE 1950S, A CULT FORMED in suburban Minnesota, led by a woman
who adopted the pseudonym Marian Keech. She predicted that during the
night of December 20, 1954, the world would come to an end, but that a
spaceship landing by her house at midnight would save her cult members.
Neither happened, and her skeptical husband slept soundly through the
night. But rather than being disheartened by this unrealized calamity, the
cult concluded that the strength of their faith had saved the world from imminent disaster and from then on went out recruiting with renewed vigor.

Early Christianity followed a similar pattern. During its first decades,
its believers assumed that the end of the world was imminent (this is clear
from the words used in the Gospels and the Book of Revelation). Yet when
the apocalypse failed to materialize, the beliefs adapted; the day of judgment became more of a metaphor, and the church was built to last, not just
to prepare for an imminent end.

The psychologist Leon Festinger used the Keech case to demonstrate his
theory of *cognitive dissonance*, which described the many ways in which we
adapt, spin, edit, and distort to maintain a coherent worldview. These ways
help us to survive and sustain our sense of self, and they do the same for
groups. But they are frequently the enemies of intelligence and, on a large
scale, of collective intelligence.

The same is true of our habitual ways of thinking. The psychologist
Karl Duncker invented the term *functional fixedness* to capture how hard
it is to solve problems, because so often we start off by seeing a situation

through the lens of just one element of the situation, which in our mind already has a fixed function. But frequently that has to be changed for the problem to be correctly interpreted, let alone correctly solved. This turns out to be particularly hard. His classic example was the "candle problem." People were given a candle, box of thumbtacks, and book of matches, and asked to fix the candle onto the wall without using any additional items. The problem could only be solved when you realized that the box containing the thumbtacks could be used as a shelf and wasn't just a container.[1]

I've already suggested that a group with a more autonomous intelligence will fare better than one with less autonomy. It will fall victim less often to the vices of confirmation bias or functional fixedness. It is more likely to see facts for what they are, interpret accurately, create usefully, or remember sharply. Knowledge will always be skewed by power and status as well as our preexisting beliefs. We seek confirmation. But these are matters of degree. We can all try to struggle with our own nature and cultivate this autonomy along with the humility to respond to intelligence. Or we can spend our lives seeking confirmation, like Keech and her followers.

Much of what is best about the modern world has been built on institutions that reinforce the autonomy of intelligence. These serve us best by not serving. They work best for their clients or partners by serving a higher purpose, and not trying too hard to keep people happy or comfortable.

As described earlier, the aviation industry is a good model of autonomy in this sense. Every airplane contains within it two black boxes that record data and conversations, and that are recovered and analyzed after disasters. Every pilot is duty bound to report near misses, which are also analyzed for messages. The net result is a far more intelligent industry and one that is vastly safer. In 1912, at the dawn of aviation, more than two-thirds of the US Army's trained pilots died in accidents. By 2015, the accident rate for the main airlines was around one crash for every eight million flights, helped by an array of institutions that study errors and disasters, and recommend steps to prevent them being repeated.

The modern world is full of institutions that reinforce the autonomy of intelligence against the temptations to illusion and self-deception. A functioning market economy depends on independent auditors assessing company accounts as accurate (and suffers when, as in the United States, the auditors have strong financial incentives to please the companies they are meant to audit). Markets depend on accountability procedures, shareholder meetings that potentially challenge managements that have become carried away, and free media that can uncover deceptions. More recently, the movement to promote open data in business has made it easier to track ownership patterns and corporate behaviors. Over eighty million businesses are tracked by OpenCorporates. These devices all exist to make lies and deception harder.

Much the same is true in governments. We can't rely on the personal ethics and integrity of our leaders, although we should prefer ones who can still spot the difference between right and wrong. As Mark Twain put it, the main reason we don't commit evil is that we lack the opportunity to do evil. So a well-functioning government depends on scrutiny and transparency that can show how money is spent, and which policies are achieving what results, all supported by bodies over which the government has limited, if any, power.

The moves to create new institutions to support evidence form part of this story, such as "What Works" centers, expert commissions, independent offices of budget responsibility, and independent central banks that have to publicly justify their decisions. All exist to make the available facts more visible so as to reduce the space for deception, delusion, and ill-conceived actions. The principle is that anyone in power has every right to ignore the evidence (since it may well be wrong). But they have no right to be ignorant of it.[2]

A healthy science system is the same. It polices itself. Thanks to peer review and tough scrutiny of research findings, the science system needs less oversight, management, or intervention from the outside. What counts as quality is transparent. There are many threats to this kind of self-governance: corporate funding with too many strings, the tendencies to suppress negative research findings because careers appear to thrive much more when research findings prove something new rather than failing to prove something, the decay of peer review, outright fraud, and hidden conflicts of interest. But there are plenty of counterforces, and in the best systems, lively debate that uses errors to make the system work better.

These defenses against deception are the corollary of the expansion of autonomous intelligence within societies and daily life—giving people the freedom to explore, think, and imagine without constraint. This was the logic of the Enlightenment as well as the great wave of tinkerers and fiddlers who energized the Industrial Revolution, and held a view of freedom as the freedom to try things out. Some of their work was logical and linear. But much of it was more iterative and exploratory: testing ideas and options for feel as well as coherence, multiplying arguments, and seeing whether they stood up or not.

ERROR AND EXPLORATION

Here we see a general feature of human intelligence: that it thinks by doing, externalizing, and then reinternalizing, in a zigzag rather than straight line, using error to find what's right. The extraordinary film of Picasso painting—*Le Mystere Picasso*—shows this in art, as the painter fills and refills the space with evolving images that first portray a coherent scene, then disrupt it, then mutate it into something new. He doesn't begin with a representation in his mind that then materializes on the canvas; instead, the representation emerges from the act of painting.

It's only through error that the right line is discovered. In the same way when we learn a new task—for example, firing a gun, playing tennis, or driving a car—we learn by going too far and then calibrating back. It's almost impossible to learn right the first time—and we risk not having internalized the lesson unless we've experienced the errors.

The best discussions in a board or committee have some of the same properties: working around a problem or options, trying out different positions, and feeling as well as analyzing which makes the most sense. It's hard to make decisions in a linear way—with a direct logic from analysis to options and then prescription. Rather, whether as individuals or groups, we need to sense the options as well as think them. The architect Christopher Alexander suggested this as a general approach to design. He liked to see how people themselves had found ways to live in the places where a building was being considered. Where did they choose to walk or sit around? Why was one corner, which caught the sun, so apparently congenial? Why did people feel so at home in one part of the piazza or park?

Then he would build in an exploratory way, trying out cardboard mock-ups of a pillar or window to see if it felt right before building it permanently. Good design couldn't be done solely on paper or a computer screen. It also had to tap into an intuitive sense of what felt right, and that sense could only be discovered and trained by showing it examples that weren't quite right, calibrating through error to a better answer. These are instances of autonomous intelligence: allowing thought free rein and following it, rather than subordinating it too easily to a single will or predetermined plan.

Intelligence, Language, and Communication

Intelligence can be autonomous to varying degrees and serves us well when it is given sufficient freedom. But we also need to consider the autonomy of the medium through which groups try to think and act together. Cognitive science has tended to see all intelligence as dependent on representation, and that trend has been encouraged by computer science, which models thought through representations. Languages aren't only tools or representations, though. They carry their own logics. Shared representations come to constitute groups, and collectives are inconceivable except as bound together by words and symbols, memories and meanings. Standardized ontologies make it possible for large-scale cooperation to happen—languages, jargons, formalized standards, or rules governing data.

But the role played by communication is never solely functional. As Friedrich Nietzsche wrote, every word is also a mask. Every ontology is selective and may reflect the worldview of its creators. And so the mediums of collective intelligence exert a strong pull on the ways in which cooperation happens, and can become the enemy of clear thought or indeed freedom.

Somehow we have to communicate to others about what we think, what we want to do, or how we might cooperate. We can point, grunt, and then talk. Yet our vocabulary will always be more limited than our thoughts and feelings, a shadow of the pure communication of perfect collective intelligence and sometimes a distortion. With communication comes degradation, error, and misinterpretation. The message sent by the sender and received by the receiver are never identical. And although a theory of mind—imagining what may be going through others' minds—is essential to cooperation, this will always be an imperfect representation.[3]

This very ambiguity can fuel deep empathy, and some of the greatest art plays on our ability to read meaning into silences and absences (as in Ernest Hemingway's famous six-word novel, "For sale: baby shoes, never worn"). But we know from the study of communication just how much all communication suffers from entropy, decay, and forgetfulness. So we rely on methods that address these head-on. They may seem the opposite of intelligence but oddly sustain it. The most important are repetition, redundancy, and rules.

Repetition is the dull heart of organization. The numbing repetition of principles, ways of working, and codes of conduct that slowly permeate into the people have their purpose. The military inculcates clear communication through relentless repetition; quartets and orchestras do the same; and many firms embed ways of thinking and procedures into their staff through repetitive training and checklists.

Every message has built-in redundancy—extra layers that appear unnecessary and show up in the checklist, rulebook, and guide. In the underlying design of the Internet's packet switching, each node sends a message to confirm it has received the message, and if this doesn't happen, it is sent again. In everyday conversation, we sometimes repeat back what we've been told to confirm that we've heard it right.

Redundancy is key to everyday language and communication. Strip messages down to the bare minimum and there is a much greater risk of misunderstanding. So words and phrases are added to pad out, and through repetition or providing greater context, these reduce the risk of misunderstanding. Rules often formalize both repetition and redundancy—rendering messages more predictable because they are more formulaic. These were the devices used in Homeric poetry to help memory, but are also used in all everyday intelligence; they are the opposite of shorthand economizing devices, but reduce the risk of error.

Epistemic Vigilance

In the worldview that sees every group and institution as akin to a computer, the main things communicated are data and commands. But in human groups, a significant role is played by the communication of arguments. Why should we go this way and not another? Why should we share the food in this manner? Why should we distrust a stranger?

Throughout our lives, we are surrounded by claims and arguments from the most minor to the most profound. This is why we pay so much attention to what Dan Sperber calls *epistemic vigilance*, asking, Can we trust this person? Are their claims coherent? This is also why we use rhetorical devices—extending from what is known or accepted to the new case, and minimizing the gap to be jumped.

There are many good illustrations of this in the psychology literature, including versions of Wason tests, which assess reasoning and logic. A classic Wason test shows cards with letters and numbers, such as *6*, *D*, *5*, and *J*, to people, and states, "If there is a vowel on one side, then there is an even number on the other side." People are then asked how they would verify the statement.

Most individuals fail these logic tests, missing the option of falsification (that is, looking at the other side of a card showing *6*). But group reasoning turns out to be a lot more reliable mainly because people argue with each other, together assessing the quality of the assertions being made. Most human groups are better at reaching answers through the clash of arguments than through joint linear logic. This takes us to an important feature of thinking in groups: they think best when they allow competing contentions to take shape, be aired, and then be judged or combined. This is not a linear deduction (which narrows the space of possibility) but rather the opposite: a way of thinking that first expands the space of options and possibilities, and then closes them down. Correct understanding takes much more time and energy than correct communication.

Still, being a group doesn't guarantee healthy debate. Status, prestige, and eloquence can substitute for clear thinking, and all groups are vulnerable to groupthink. As I will show later, collectives with any sense of a *we* or subjecthood, like a successful business, charity, political movement, or club, always create their own languages and codes, which then simultaneously help to define the boundaries of the collective and produce a logic that is in part autonomous from the group that the intelligence is meant to serve.

The same is true of the great diversity of specialized languages that allow for coordination on much larger scales, the structured languages of bits and bytes, or bar codes and URLs, and financial or medical languages. These often embody not just an accumulated knowledge but also an implicit worldview. They think through us as well as us through them. They serve precise functions, but the language can then come into tension with the group that it's meant to serve, since it limits and constrains, and defines out and in.

The languages we use to communicate and think with also contribute to one of the other vices of autonomous intelligence: what could be called the law of overextension. It's a feature of the human mind that we tend to extend ideas from one field to another. This contributes to a great deal of fertile innovation: taking ideas from the school into the business, from airports into hospitals, or from play into learning. But this tendency also overshoots, so that every answer that works in one setting risks being extended too far, losing sight of its moment of origin. The fight to protect from danger becomes a habit that persists when the danger has passed, and in extreme cases leads to wholly militarized societies. The pursuit of money that was once a tool toward a better life becomes an all-consuming obsession. Language makes it easy to transpose and translate. At its best, this allows us to think laterally and flexibly. At its worst, it encourages cascades of "category error," as we use the same words and modes of thought for what are in fact different things. As a result, every means becomes an end and every escape becomes a trap. So as intelligence becomes autonomous, it can be both beautiful and ugly.

The World Wide Web is, perhaps, the ultimate example. Every piece of information has a Uniform Resource Identifier, which means that the map is also the territory.[4] Representation is explicit, shared, and designed in and through social media ever more tightly coupled with individual intelligence. Here language is freed from ownership by big governments or companies, religions or movements. It appears universal, objective, and open. Yet here too representations can be autonomous, freed from any objects of representation, and so can become a threat to intelligence. Here too is a sphere in which lies, fantasies, utopias, and false dawns can spread without limit in the internal logic of the web. Here too an individual can construct an identity wholly different from their identity in the real world—as has happened repeatedly with terrorists.

There's a comparable struggle within us all. Our bodies' metabolism regulates the heartbeat, blood pressure, breathing, and digestion, the ins and outs of food and waste, and oxygen and carbon dioxide. This metabolism supports cognition, but the ideas our brain produces can move ever further from their metabolic foundations—giving us baroque religions, fantastic arts, and deluded fantasies. Cognition can get carried away and create coherent worlds that bear little or no relation to the one we live in.

Again and again, though, the material world pulls them back, as happens each morning when we wake up, needing something to eat. The imagined world can clash with the real one. God may not appear on schedule to provide a timely miracle. A market doesn't behave as predicted in the textbook. A lover doesn't follow the script of the romantic novel. The meanings are partly autonomous, but not wholly. The flower has its roots in the mud. And so we learn to rein intelligence in—to allow it freedom, but a conditional one that is regularly tested and attached to the world it helps us navigate.

- 9 -

The Collective in Collective Intelligence

I did not get my picture of the world by satisfying myself of
its correctness: nor do I have it because I am satisfied of its
correctness. No: it is the inherited background against which I
distinguish between true and false.
                              —*Ludwig Wittgenstein, On Certainty*

WHAT IS THE COLLECTIVE in collective intelligence? What makes a disparate group of individuals become something more like a *we*, as in the
example of the stranded passengers left at an airport?

All around us are groups and organizations that have some of these
properties. A collective personality may be embodied in a firm, university,
political party, or government. We rely on many of these to answer basic
needs, for health care, mobility, security, or faith. So how do they think,
and how could they think better?

For Descartes, it was evident that cogito, ergo sum—I think, therefore
I am. But does that apply to a group? If a group thinks, does it therefore
become an *I*, a subject? Evidently not, as many instances of large-scale
thought don't create subjects in any meaningful sense. The big collabora-
tions online—like Wikipedia or GitHub—don't require the people taking
part to know each other or feel any great affinity. Markets can adjust prices
without traders needing to feel any emotional bonds with other traders.
We can be troubled by events on the other side of the world without feeling commitment to the people affected (as one writer put it, in the age of
the Internet, "a trouble shared is not a trouble halved. It is a trouble need-
lessly multiplied all over the world").[1]

Conversely, the intelligence of collectives doesn't necessarily depend on
the intelligence of the individuals that make them up. Eusocial insects,
marmosets, and many other creatures have learned to share goods, send

signals, and collaborate together without much in the way of individual intelligence.

But the degree of *we-ness* matters for the everyday life of collective intelligence and how much work people are willing to put in, how much they are willing to share scarce knowledge or information, and how willing they are to stay loyal through difficult times.

Collective Consciousness

Let me start with the question of *whether* a group can think in any meaningful sense. There is little doubt that it can have purposes and intentions, even if individuals within the group don't fully share these purposes.[2] But there is much less agreement about whether it can be conscious in the sense that an individual can be. Can it have a headache? Can it experience fear and love? Groups certainly can have agency—beliefs about the world, wants and wishes, and a capacity to act. Consciousness, however, is trickier to pin down. In our everyday use of the word, it includes some awareness of things as well as the experience of being in the world. Groups can be aware of things—a police force can be aware of evidence or a political party aware of public opinion. And they can be asleep or awake; many organizations effectively fall asleep at night and during holidays.

Yet it's much harder to show definitively and meaningfully that they experience something comparable to the individual experience of being conscious, of feeling light and dark, hunger and thirst. Indeed, we can be fairly certain that individuals within the group will either disown or reject the beliefs of the group they're in. And so we are left unsure whether it is meaningful to talk of General Electric, Samsung, the Red Cross, or the nation of Belgium having consciousness.

Thomas Nagel wrote an influential philosophical paper titled "What Is It Like to Be a Bat?" His point was that conscious beings have a distinctive experience, such that there is something that it is like to be that thing, whether that is an elephant or gnat. His purpose was partly to contrast rocks, chairs, or cars that are not conscious, such that there is nothing it is like to be them. But he also wanted to distinguish the functional elements of thinking from this additional element that couldn't be reduced to functions.[3]

The same may be true of groups and how we think about their consciousness matters. It matters for law and morality. Can we meaningfully talk about the guilt of a corporation or nation? Is it meaningful to talk of a group having rights?

In many legal systems there are positive answers to these questions. A company that has caused harm can be held accountable for that harm.[4] So far, so reasonable, particularly when that responsibility is also made individual for the executives and directors most directly involved in the decision.

But this idea can be taken much further. In the United States, for example, the Supreme Court gave lobby groups the same rights of free speech as individuals. Nations have been held responsible for crimes committed many generations ago, from slavery and famine to genocide. These cases show why we need more clarity about the relationship between collective or group agency and group consciousness, which might in turn lead to rights, on the one hand, and obligations, on the other.

I want to suggest that we can find an answer in some of the ideas already covered in earlier chapters. Consciousness is both a matter of degree, the extent to which the various dimensions of intelligence are in play, and a matter of integration. The human mind is highly integrated. What is called *integrated information theory* posits the reasonable idea that consciousness is a phenomenon of highly integrated thinking. Evidence from brain scans indicates that we wake up when our brain's capacity for information integration wakes up and we fall asleep when it turns off.[5] Such an integrated system has feedback as well as feedforward, with recurrent connections across its many ways of thinking. This is much of what is being done by our eighty-six billion neurons and their hundred trillion interconnections.

No group we know has anything like the degree of integration of the human brain. So there is always an asymmetry between individuals and groups. The individual is more coherent, self-contained, and integrated than any group can be. They have persistence over time and space, memory, and consciousness of decisions and meanings. Even the most persistent group is more partial than this. Its decision makers change; its memory depends more on artifacts and records.

This is what surely justifies the view that only individuals should be treated as fully moral agents. A group can have agency and should be treated in law as such. But it doesn't and can't have rights analogous to an individual consciousness until and unless it can demonstrate a comparable level of integration. By the same token, it is a category error to call a nation to account for something done many generations ago. The integration of the leaders and peoples of the past with today's citizens is too attenuated for this to be meaningful.

The Study of We-ness

So a group is a we, but it is not just a scaling up of an I. How, then, should we think about the character of this we?

This question is missing from much of the literature on collective intelligence, which is really about aggregations of individual intelligence solving problems or editing Wikipedia entries, as opposed to the kinds of groups or organizations that dominate the world around us.

These aggregation devices can be extremely useful. Yet their limitation is simply that they aggregate, and more doesn't always mean better. Having one million ideas may not be better than having a hundred if all the ideas are of poor quality. Having ten million inputs from consumers about their preferences may be less useful than a carefully selected and representative thousand, just as a research study on a large data set may be much less useful than a more precisely designed study on a smaller group. Similarly, aggregating lots of views isn't a good way to devise a strategy or new idea, make subtle judgments, or solve "highly dimensional" problems. None of the aggregation platforms has yet played much part in solving a truly important challenge. For these, there usually needs to be some kind of structure and invariably some kind of collective: an institution or group that is conscious of itself and has boundaries.

Our understanding of we-ness is hampered by the influence of powerful intellectual traditions that struggle with any notion that a collective mind could be more than the sum of its parts. These traditions were an understandable reaction against vague and mystical invocations of community, god, or national spirit in the past. As a result, large areas of social science see collective intelligence as nothing more than the aggregation of individual intelligences, and this "methodological individualism" dominates not just modern economics but also much research in psychology and computer science. For these disciplines, any concept of a group mind is woolly and abstract. In any case, every group mind manifests itself in individual minds, so why not make the individual mind the unit of analysis?

This view still persists. But it has become less tenable. Instead, a more nuanced position has emerged that neither exaggerates the individuality of individuals nor assumes a pure group mind. It sees individuals as shaped by groups, and groups as made up of individuals. This view is helped by the ways in which psychology and neuroscience have revealed that the individual mind is better understood not as a monolithic hierarchy, with a single will, but rather as a network of semiautonomous cells that sometimes collaborate and sometimes compete. If you accept that view, then it becomes more reasonable to see groups in a similar way, even if you differentiate the highly integrated individual mind from the less integrated group mind (in other words, not altogether integrated individual minds not altogether integrated into larger groups).[6]

DISPOSITIONS

Why are we able to become part of larger groups that make claims on us? The answers lie in our makeup and dispositions. We are well designed to become part of a group and learn to do this from an early age—how to be we as well as me. We do this in play, sports, or singing together, or within a family. It's part of the repertoire of being human.[7]

Indeed if Michael Tomasello is right, this is what most distinguishes humans from the great apes.[8] The apes are essentially individualistic. Even when they cooperate (for example, when foraging for food), a group of chimpanzees that finds a tree full of fruit will each help themselves to the best food they can find. Humans, by contrast, more often support each other, make group decisions, and share, alongside the usual pushing and shoving.

The evolutionary pressures to share in this way are obvious. Faced with a hostile climate or threatening beasts, cooperators are likely to survive better through hard times or discover new hunting grounds more easily. If they can share information or manage a division of labor, they are more likely to thrive, and that will of course depend on each individual being able to take on the perspective of the larger group.

Game theory offers a useful tool for thinking about this and perhaps a more realistic one than the prisoner's dilemma. Imagine two hunters out hunting. Hunting alone, each might be able to catch smaller game, like a hare. But if they can collaborate, they have a much better chance of catching a high-calorie stag. To cooperate, they each need some sense of how the other hunter thinks, a shared sense of the purpose of their cooperation, and although they may spend much of their time hunting in parallel, as soon as one spots a stag they have to switch quickly into a we mode, acting as one and collaborating for the kill.[9]

There are many similar examples that explain why both cooperation and empathy might have spread, and as a result, why so many people can juggle so many different collective identities through the course of a typical day, as parts of families, firms, communities, interest groups, and friendship networks, calibrating in each case how much they can claim of us and how much our thought needs to be subordinated to that of the larger whole.[10]

These groups follow patterns that are not ironclad laws but nevertheless are surprisingly common. Robin Dunbar has argued that there is a rough mathematics of commitment. Forty percent of our social interactions happen with 5 other people, and 60 percent with only 15 others. It's hard to have a sustained conversation with more than 4 people. It's hard for an intensely committed group to be much more than 12 in number. And somewhere around 150 is a typical upper size for a close-knit community.

To be part of a group, we have to be able to understand how others are thinking. This is where what psychology calls *theories of mind* flow into any discussion of collective intentionality. We learn how to guess others thoughts; we learn both direct and indirect reciprocity, the principle that if you scratch my back, I'll scratch someone else's; and we learn how to read reputation and trustworthiness in others.[11]

Forging a Collective: Codes, Roles, and Rules

So if our dispositions make us well suited to become part of collectives, what then turns a disparate group into a collective—one able to think and act coherently, with some degree of integration? What turns a group of strangers into a we?

It's possible to be quite precise about the answers. The group of strangers thrown together in an abandoned airport points to many of the answers. For them, as for any group needing to become a collective, the first steps may involve improvising ways of communicating with each other and some rough-and-ready rules.

These patterns are common, and any kind of collective with the capacity to think and act as a we will draw on three essential elements that help it to come closer to the integration I described earlier, even if no groups achieve the same degree of integration that the human brain manages.

First, the group needs codes. Every group has its own language. Indeed, you could say that the very definition of a strong group is that it generates distinct codes that are opaque to nonmembers. These help to bind the group together and also differentiate it. They reduce the "semantic noise" that otherwise ensures that the hearer hears something quite different from the meanings intended by the speaker. At first they may be rough-and-ready creoles; in time they evolve into full-fledged languages.[12]

Next, any group needs roles, allocating tasks and functions to people and things. For example, a group might describe a spherical piece of rubber as a ball or identify one of the citizens as a sheriff. These are assertions that become objective through use and repetition. We do this automatically, from children's play to parliaments and monies, and find it easy to express the kind of formula suggested by the philosopher John Searle: "X counts as Y in context Z." For instance, such a person is the leader in this context (but not others). The group defines the roles, but the definition of the roles also defines the group.

Finally, the group needs rules that can govern the interaction of things and people. Some of these rules are formal, and may be set down in a rulebook, law, or charter, while others are more informal, embedded in the mores of the group.

A shared structure of this kind, consisting of codes, roles, and rules, can be found in any collective, from programmers working on Linux software to the Chinese People's Liberation Army, from General Electric to the Red Cross. They are economizing tools that reduce the time and cognitive energy needed for groups to function. They help to automate and streamline what is otherwise hard work, and so make good sense in terms of cognitive economics. Each is a logic of representation—of things, people, and processes or actions—that makes thought possible, and it can be helped by artifacts—books, films, or such things as traffic lights—that give us common patterns of thought and, again, reduce the need for hard work to communicate as well as explain.[13]

How they function is complex. The university is a federation of groups and networks, each with its own identity—they can be disciplines, colleges, or cliques of collaborators—and each of which can switch to a different we depending on the context. The firm is likewise made up of units and divisions—the vice of silos is also the virtue of commitment and mutual bonding. The army is made up of regiments, units, and divisions that make a first claim on loyalty. But these many units learn to fit too, however uncomfortably, into the larger unit, a we made up of many we's.

The surprising pattern in all these cases is the interweaving of the sense of we-ness, artifacts, and the ability to think as one. A good example of this is the story of what happened in 1952 when firefighters were dispatched to contain a forest fire on the upper Missouri River. Disaster struck when a group of the firefighters misread an out-of-control part of the fire and died trying to escape.[14] The fire had jumped across a gully. The leader of the group had realized the gravity of the situation and ordered the crew to throw away its tools. But by doing so, he weakened his authority, and when he told them to move into his escape fire (an area of land he deliberately burned so that by the time the main fire arrived it would be safe), they no longer recognized the authority behind it. Only two survived.

The episode became the subject of a book and fascinating study by Karl Weick. He showed that panic didn't cause the group to disintegrate but instead resulted from it. When the group lost its defining artifacts, it fell apart. The social order helped to keep the cosmological, epistemological order in place. Yet when it fractured, the sense making fractured too.

Shared Models to Think With

We saw earlier that the model is the starting point for individual intelligence, coming before data and observations. Human brains construct models of the world and think through them.[15] Inputs of data are refracted through models—which are tied to a sense of being a self in time and space. Our brain compares new inputs against the assumptions of our models, and in this sense we experience the world indirectly, not directly.

There are good reasons for believing that collective intelligence in its more developed forms is similar. It can emerge from the interactions of many individual parts, without any mutual understanding or commitment—whether in markets or communities solving problems in mathematics. But the more dense the model, the more it seeks to explain, the stronger the sense of we, and the more it travels toward an ideal of integration.

The codes described above do some of this work. They are just the starting point, though. A strong collective will also have a shared model of how the world works and how things happen. That, too, can be a powerful economizing tool, since it helps the members of the group think much more quickly together. Managers at Amazon, for example, will have absorbed a view of the world as made up of data, an assumption that new ideas should be tested experimentally, an ethos that favors growth as a good in itself, and weak dispositions to attend to information about feelings, mental discomfort, and many other daily human experiences. Doctors working at Médecins Sans Frontières will have a different model, which attends to diseases and injuries, applies clinical models for treating them, and is guided by an ethos of service that is almost deliberately blind to the conflicts swirling around them.

These models can be thought of as agreements on how to define and discover truth. They are agreements about "what is"—how the world works—and "what matters"—and therefore is worthy of attention and action. They are the two dimensions of any strong collective intelligence (with what matters reaching into the relevant past and possible future).

The first half of this—the view of what is and how the world works—can become a field for science, data, and testing (and in chapter 10, I look at how collectives orchestrate self-suspicion to improve their understanding of what is). What matters, by contrast, is more endogenous and shaped by the group.

These models can, like the codes, become tools for conflict as well as cooperation. Anthropology has often emphasized how much groups are defined by their enemies, the binary view of the world as divided into us against them—we, the pure, against they, the impure and disgusting; friends against foe. If people aren't made to feel included, they can turn from friend to foe ("if you don't initiate the young men into the village, they will burn it down just to feel the heat").

These oppositions strengthen our commitment to our own group, and allow us more easily to feel both collective pride and collective shame and guilt. Sometimes the best way to strengthen a group is to attack it—which is why bombing civilians has so frequently backfired as a military tactic, and why demagogues and dictators who are losing support so often seek out a conflict, real or imagined.

The more extreme examples of we-ness verge on the unthinking, such as a troop of soldiers instinctively responding in the same way to a sudden attack, or an orchestra or jazz band, scientific team, or political campaign. There is an impressive aspect to this, as in a sports team where each member almost mysteriously guesses how the others will respond or where the ball will be kicked, or a jazz band improvising. This is collective intelligence as mutual sensitivity.

A different kind of mindlessness can be found in the cult or squad that has so suspended a sense of self that it will do anything the group or leader demands. In these cases, the model overrides active intelligence. We speak of the disciplined marching band or group of infantry trained to respond automatically to a threat as being regimented. This is a particular kind of group mind that is literally embodied and can respond in a mindless way because formulaic responses have been learned so intensively. It is an efficient way of economizing on cognitive resources, but unappealing precisely because it requires the individual to give up so much freedom.

Teams

There is a huge literature on teams and groups and the dynamics of decision making, dating back to Marjorie Shaw's classic 1932 study of group reasoning that showed that teams are better at generating options and correcting errors of reasoning. Subsequent research has confirmed that teams are usually better at making decisions as well as finding information and assessing people, despite some exceptions.[16] Even burglars who work in groups face less risk of being caught than those who work alone.[17]

Research on what makes teams great, whether in sports or the military, shows that although it helps to have outstanding individuals, many teams are far better than their parts. This happens where there is some sense of shared purpose, or a shared view of standards and quality—what counts as good, a shared ability to focus, an appreciation of the contribution of each member, and perhaps most of all, a serious and even brutal honesty when things don't go right, thereby fueling a culture of learning.[18]

In other words, teams have their codes, roles, and rules, but also go further to create a living model of the world that they can constantly interrogate and improve. The best teams allow their intelligence a degree of autonomy from the interests of each member and so serve the interests of the team as a whole. They create a commons, and the qualities of that commons determine how well the team does.[19]

This isn't the whole picture. Groups can amplify sins of commission, omission, and imprecision, and groups can make riskier decisions.[20] But despite its flaws, the team that is more than the sum of its parts and thinks like a team, not just a collection of individuals, continues to be one of the most important units in almost every field, from warfare to sports and business.

Collectives and Societies

In any society, there are many more collectives being born than actually survive. These may be groups of friends, business start-ups, or aspiring community groups. All begin with someone or some people trying to persuade others to adopt their worldview, to "buy in" to their way of seeing things along with what's desirable and possible. We can fairly easily slip into the mind frame of others, for the reasons cited above. But we can equally easily slip out again too, if we don't see enough interest, meaning, or fun in playing along. Many collectives are deliberately set up as temporary—to fulfill a particular task and then disappear.

These patterns can be found at every level, as groups compete to attract adherents. At the macro level, the central insight of much nineteenth-century sociology was that how we think reflects how our societies are organized. As people interact and try to make sense of each other, they generate shared categories. The ones that take hold tend to reflect their social order, and so everyday conversation confirms that order, at least in the first instance. To that extent, the larger unit thinks through us rather than the other way around.

Alexis de Tocqueville wrote sharply about how this happened in nineteenth-century America. We tend to conform because we fear the disapproval of our peers or authority—the theme of Hans Christian Andersen's "The Emperor's New Clothes." This can lead to optical illusions. As old beliefs wane, "the majority no longer believes, but it still appears to believe and this hollow ghost of public opinion is enough to chill the blood of would-be innovators and reduce them to respectful silence."[21] Often people believe that the majority of others believe things that they don't (for example, in Israel nearly 60 percent agree to Palestinian autonomy in the territories, but only 30 percent realize that this view is held by a majority). Such optical distortions—called *pluralistic ignorance* by sociologists—can have a great effect on how groups think, while also helping to reinforce the status quo.

The connections between macrocultures and beliefs have been studied in detail through history, and certainly confirm the extent to which collectives think through individuals. One of the best examples is, ironically, Peter Laslett's study of how the nuclear families of England in the Middle Ages helped usher in the market individualism of the Industrial Revolution. Emmanuel Todd generalized this approach, showing how much deeply rooted family structures shape how we see the world and determine which political ideologies resonate most because they seem most natural, fitting with the world as we've grown up with it. Todd's work explained which areas of Europe were most communist or nationalist. Family structures correlated with the dominant ideologies and seemed to cause them too.[22] Much of sociology has been dedicated to disentangling these effects of larger units on individual thinking and behavior.[23]

How Different Collectives Think

These rather-abstract theories about how societies think can be made more practical and visible in everyday life if we turn to how smaller units think. It will already be clear that there are different types of collective. The cooperation of shopkeepers accepting the same credit card has little in common with the cooperation of a troop of soldiers, or group of software programmers and artists. So we find divergent patterns of types of collectives, each of which thinks in a different way, and not all of which require a true "meeting of the minds."[24]

A useful way to map how groups think is to distinguish them according to two axes (an approach first pioneered by the anthropologist Mary Douglas). One describes collectives according to their level of *grid*: how much formal vertical hierarchy, control, and authority there is. On the other axis is their degree of *group*: how much horizontal bonding and mutual commitment there is. This provides a two-dimensional space within which to place such different phenomena as Wikipedia and citizen science (very low grid and fairly low group), transactions in a marketplace (very low grid and group), the multinational corporation or military (high grid and sometimes high group, but often high grid and low group when the soldier or worker fatalistically goes along with their orders even though they have little real commitment to the institution they're part of). These provide distinct answers to the questions that define any group: What is, and what matters?

In low grid and low group cases, the mechanisms of invisible hands can make a large group behave as if it were a collective intelligence, at least in a small number of dimensions. Widely dispersed markets can use price signals efficiently to coordinate activity on a huge scale (though as we will see, all real economies also depend on different types of collective intelligence). The collaborators solving a math problem or advising a chess player online need not know each other, or have any common identity. There need to be simple enough common grammars and languages, and straightforward rules of the game. But with the right rules, quite large and complex activities can be coordinated with light collectives that have no sense of self.

Where group commitment is low, incentives and financial rewards can be used to encourage people to align their behaviors. People are, in effect, bribed to stay on the team and cooperate with colleagues who they may neither like nor identify with. Some aspects of competitive scientific research are similar. The utopia of this worldview is the perfect market where the invisible hand aligns many actors.

These types of collectives, however, are much less effective for handling force, adapting to threats, or mobilizing long-term resources. They can't make claims, demand sacrifices, or resolve conflicts. That's why many functions in fields involving security, long-term investment, or health tend to be dominated by groups or institutions that are higher in both grid and group, which can use coercion or the threat of dismissal, and can reward people with recognition as much as money. Power and responsibility are organized into chains of command, and any breaking off from the group is seen as a kind of betrayal. This is the logic of the typical state, corporation, army, or traditional family, all of which are high grid and group.

For the high group and low grid communities, the rewards come from love, solidarity, and mutual care. These tend to be egalitarian, and their ideal is the small group that's flat in structure, reaching decisions by consensus, and their utopia is the perfect community, where everyone is closely attuned to the needs, wants, and emotions of everyone else.

In these different ways, the various types of collectives align behaviors, but they also help people to think alike. The individualist perspective interprets the world through the lens of interests and incentives, the hierarchical one through conflicting power, and the egalitarian one through the self-organization of the people. Each culture also has its own communications logic. Egalitarianism is rich in communication, yet sometimes impeded by the costs and slowness of high-bandwidth communication (which is simpler in small groups than large ones and for slow rather than fast decisions). Hierarchy has more specified channels for communications upward and downward, with less horizontal communication, which allows for speed and less ambiguity. Individualism works best with simple currencies for horizontal communication and reward (for instance, market signals or scientific prestige).

All three of these cultures can be found coexisting on the Internet. There is the counterculture egalitarianism of hackers and the open-source movement. There is the aggressive commercialism of the big providers, funneling returns back to venture capitalists. And there is the hierarchy of the military, which funded the early Internet, and thinks in terms of security threats and geopolitical competition. Each of these cultures is a tool for thinking, which prioritizes some facts relative to others, and each clusters around particular models of the world. These ways of thinking help them make decisions, but they also help them cooperate, since individuals within these cultures absorb them, becoming useful conformists.

Indeed, even cultures that think of themselves as individualistic, dissident, and rebellious tend to be highly conformist. The beliefs held in a Silicon Valley conference or meeting on the Left Bank in Paris are not much less conformist than those of a religious seminar or the military.

But the crucial point is that these diverse ways of thinking are complementary as well as competitive. This was the surprising insight of grid-group *culture theory*: that within any group, these cultures will all be found even though one is likely to be dominant. Every hierarchy requires some tools to incentivize and reward individuals, and in all likelihood a grass-roots egalitarian counterculture (the troops in an army or the shop floor in a factory). Every market needs hierarchies within it. And even the most egalitarian movements turn out to need some hierarchy to make decisions and some incentives to keep the bread being made.

The most advanced forms of large-scale collective intelligence therefore combine each of these cultures, allowing for a pluralism of worldviews and models that will always be to some extent in competition with each other.

In chapter 5, I described quite different kinds of collective intelligence in the wild: the group of passengers stranded in an airport, the world grappling with climate change, and a small garage in a small town. If we look closely at each of these, we're highly likely to find just such hybrids of different cultures, and in each case, if one culture becomes too dominant, mistakes are more likely to be made.

Climate change is a good illustration. There is no shortage of advocates arguing that only one kind of solution will really work in reducing human emissions. For some it is obvious that this has to come from hierarchy along with the imposition of stringent taxes and controls, preferably through international treaties. For another group it is equally obvious that the answers have to come from the market or the natural emergence of new technologies. For a third group, the only answers lie with communities themselves having a change of heart.

All are wrong. Yet all are also right. It will be the combination of these three distinctly different models of the world that will lead to successful collective action. The same is true of other assemblies, like the world of medical knowledge, which combines strong hierarchies (in medical research, hospitals, and professional bodies), strong horizontal markets supporting the creation and sale of drugs as well as instruments; and a strong egalitarian ethos among millions of doctors and nurses. This is the insight that comes from understanding how different groups think: they complement, compete, and interact with each other in ways that mirror the nature of a world that cannot be captured with just one way of thinking.

THE PERSISTENCE OF COLLECTIVES

So what are the advantages of having a strong sense of collective self—and what are the disadvantages? Why might organizations be better at handling complex problem-solving tasks than looser aggregations?

Obvious answers include their superior ability to mobilize more re-sources as well as sustain the commitment and persistence that's needed for difficult, long-term tasks. But I suspect that the payoffs go deeper than this. The sense of a collective self may also make it easier to make subtle judgments, and mobilize second- and third-order loops of learning than more distributed systems. Second- and third-loop learning are both hard work, and require a more integrated intelligence. In a fully decentralized system, each unit makes these judgments for itself, exiting from links that are no longer useful and adopting categories from others. Aggregation is easy; integration harder. For a collective with a sense of self, the task of generating new categories is shared and perhaps more explicit, such as a team that is no longer winning its matches, a city that is stagnating, and a company losing its market share. This will be even more the case if there's a need for third-loop learning and a shift to an entirely different way of thinking. Without this kind of grounding, information always risks re-maining meaningless—stuck in the quantitative Shannonite world of bits and information flows.

We can see a version of this problem in the United Nations and global bodies. To the degree that the United Nations is only an aggregation of nations, occasionally cooperating when it suits them, it struggles to rein-vent itself as a true collective intelligence would. It can handle first-order learning within existing rules and models. But it is not well designed to invent new categories or acknowledge when the old models are no longer working. And it is even less well suited for third-loop learning, reinvent-ing itself as a system of intelligence. Because any change to a new system would involve losers as well as winners, its early structures are frozen. New categories have to be invented outside the system rather than within it. Again, it is good at aggregating yet not so good at integrating.

Other reasons have to do with uncertainty and change. Strong collec-tives can combine the crucial capabilities needed to cope with a radically altered environment. These include a shared view of the future along with an ability to move resources from one task to another to cope with new priorities. When these exist, the result is much greater resilience and adapt-ability. Overly rigid hierarchies are not so good at helping a group mind adapt. But distributed networks are also poor at this kind of adaptation. Like the United Nations or other loose federations, distributed networks struggle to reallocate resources—one of many reasons why decentralized guerrilla bands find it so hard to win wars and highly competitive markets struggle to pool sufficient resources to develop scientific breakthroughs.

A group that really can coordinate and align actions so as to act as a strong collective intelligence is likely to be more successful than one that cannot. Many tasks require intense mutual coordination and communication— when one element of the network changes, others need to change too (think, for example, of an army fighting a battle). This, the interdependence of many tasks, explains the limits of the wisdom of crowds. Crowds can estimate stock prices, the numbers of beans in a jar, or the weight of a cow, as Francis Galton famously showed in the late nineteenth century.[25] But you would be unwise to hope too much for the ability of any crowd to create a memorable work of art or write a book that anyone would want to read—tasks that require much higher bandwidth and iterative, mutual communication, or again, integration rather than just aggregation. Markets can adjust prices with little horizontal communication between the participants, other than the binary decision of whether or not to buy. Redesigning an entire supply chain, by contrast, requires high-bandwidth mutual communication and understanding, and similarly, many creative tasks turn out to require teams to build up significant levels of mutual understanding and empathy, which is why face-to-face interaction is much more effective than online interaction.

The same is true of commons. The work of Elinor Ostrom and others on commons, which is essentially about how, with rich communication, groups can manage common resources, confirms the more active and interactive ways in which they work. Indeed, it is the very richness of their communication that marks them out, as intelligence is organized in polycentric ways between multiple centers needing to collaborate over common resources or tasks. Markets relying on financial incentives and hierarchies reliant on authority are insufficiently rich in bandwidth to cope with the constant adjustments and redesigns needed for commons management.

The importance of richer, more intensive communication based on shared models, codes, roles, and rules helps to explain why—a generation after the invention of the Internet, and after decades of predictions that an era of distributed systems would bring the end of the large corporation or big government—these and many other kinds of institutions that demand loyalty continue to thrive. Government shares of world GDP are higher, not lower, than thirty years ago, just as the share of the largest corporations in the global economy is bigger, not smaller, than it was. Groups with more we qualities are more resilient, and thus better able to cope with setbacks and defeats.

These points don't matter only for organizations or whole societies. They may also turn out to be decisive in explaining why some places do so much better than others. Influential research by Felton Earls, drawing on studies in Tanzania and Chicago, showed that what he called *collective efficacy* is critical to explaining crime rates. The most important influence on a neighborhood's crime rate is neighbors' willingness to act, when needed, for one another's benefit, and particularly for the benefit of one another's children. His detailed research rebuffed the fashionable "broken windows theory," which claimed that often-superficial signs of order were crucial in encouraging more serious crimes.[26] Its implication was that cities should encourage neighborhoods to work together to solve problems, learning by doing to create a sense of collective efficacy and intelligence that would then discourage crime and antisocial behavior.

SCALE AND COLLECTIVE THOUGHT

What effect does scale have on the dynamics of collective intelligence? In the natural world, scale affects structure and processes in myriad ways. A heavy animal needs a very different structure than a light one. Weight increases as the cube of height—hence the need for proportionately much larger legs. This is why insects are so different in shape from people, who are in turn so different from elephants. Surface area increases as the square of height—hence the need that larger creatures have for much more developed systems for circulating blood or cooling. Viscosity changes the shapes available for small things—which are closer to the size of molecules, and so can stick to walls (like gecko's feet) or the surface of ponds (like backswimmers).

These patterns are well attested to in animals and plants that are far from being fractal. But what equivalents are there in human institutions, or ones combining humans and machines? Is a bigger church, firm, political party, or government more intelligent than a smaller one, and if so, why? Or is the opposite true?

Economics suggests largely linear economies of scale. The more cars or aircraft you produce, the lower your costs, given what you have learned, because you can purchase cheaper raw materials, keep lower inventories as a proportion of output, and so on. Larger markets will tend to be more efficient than smaller ones, mainly because they can generate more economies of scale.

But there are many fields where these patterns don't show up. The best schools aren't the biggest. Instead, they are about the same size as other schools. Nor are the best care homes or hospitals the biggest ones. Anything involving love and care seems to thrive more on smaller scales, and making complex decisions seems to be harder with large numbers, even if it can be helped by observation and commentary from many.

In government, I once commissioned a study of economies of scale in public services. It was assumed that these were substantial, and every few years a management consultancy recommended consolidating power into larger units so as to save money and allow more specialization. To our surprise, we found almost no evidence for this. The same services could be efficiently provided with a unit of fifty thousand or five hundred thousand (though the smallest and largest units tended to come out worst). The same finding has come from research on governments that can be highly competent with three hundred thousand citizens (like Iceland), thirty million (like Malaysia), or three hundred million (like the United States). In this sense, what governments do is fairly fractal in nature. The advantages conferred by scale in terms of geopolitics or economics might appear to be balanced by the greater risks of delusion and capture by special interests.

Yet some other features of scale more closely mirror those in the natural world. Comparing big nations to smaller ones, the greater size, weight, and surface area of the big ones require different tools that echo what's found in nature. The big ones need more law and governance for every element. There's more use of devices to circulate the ideas of the top—compulsory national curriculums, flags, patriotic devices, broadcasting, party units, and cells. There may be more paranoia because of the surface area and a larger number of potential threats relative to size. There may be more abstraction in defining ideas relative to the village-scale norms of small nations.

Certainly, as institutions grow, their forms change, with a tendency to create more structure, bureaucracy, and formalization. Roles become more impersonal and are more separated from the individuals carrying them out. Cultures are made more formal rather than tacit, with more explicit incentives as well as more hierarchy.[27]

There are also benefits of scale: more resources for intelligence, knowledge, specialized skills, experience, and access to networks. It's easier to formalize second- and third-loop learning—for example, with specialized strategy teams, analysts, reviews, or buying-in consultancy.

And if power is defined as the freedom to make mistakes, size appears to give more freedom. Yet large scale also corrodes the very qualities that make for judgment, like awareness of context and self-suspicion. Large entities are more at risk of becoming trapped in their own abstractions and believing their own myths

Small groups can think things through with conversation or trial and error.[28] But on any large scale we need things to think with, and artifacts play a vital role in turning tacit codes and rules into explicit ones, such as books, websites, and symbolic objects like flags and uniforms, rendering the groups' tacit knowledge visible. Indeed, it's possible that the larger the entity, the more it will rely on artifacts. Empires are littered with statues and memorials. The United States salutes the flag. Estonia perhaps doesn't need to. Big political parties and businesses work harder on their logos, their synthesized identities, than small ones.

As we have seen, everything from standardized sizes for cars to bar codes, writing scripts to plate sizes, assist in the everyday work of cooperation, and these work best on a large scale, creating new social facts that help us to get along. Yet the most important fuel for collective intelligence comes on much smaller scales, such as bodily propinquity, eye contact, and mutual feeling. This is where the theory of mind becomes functional. We think our way into the mind of the team we work with or the board members we sit with. But there are no reliable ways of scaling this insight. This explains the survival of small groups, boards, committees, and cells as units of action and decision making. Mutual awareness of mind and then the creation of a common mind seem to be bounded by scale and the limits of emotional bandwidth in the human brain—the limits on just how many people we can feel with.

- 1 0 -

Self-Suspicion and Fighting the
Enemies of Collective Intelligence

The greatest obstacle to knowledge is not ignorance; it is the
illusion of knowledge.

—Daniel J. Boorstin

A FAMOUS SERMON BY THE BUDDHA, the Kalama Sutta, sets out ideas that
run counter to almost every religious text of the last few thousand years.
While other texts asserted claims about the world and cosmos, the Buddha
advised his listeners, a tribal clan called the Kalama, to think critically: "Do
not act on what you've heard many times; or on tradition, rumor, scrip-
ture, surmise, axiom, specious reasoning, bias, or deference to monks." He
was warning his listeners to be skeptical of itinerant preachers like himself
and know that the road to insight was paved with doubt.

His worry was not that his listeners were stupid but rather that the very
qualities that make us intelligent also make us foolish. We can be trapped
by habit and deference. We can be lulled into a false sense of confidence
by familiar words and notions, which means that we stick with a way of
thinking even though the environment has changed. We can be misled by
our tendency to take an example or anecdote as more representative than it
really is. It is easy to be fooled by an apparently charming and honest con
artist, conspiracy theory, or vivid story.

Equally, we can be trapped by our own ability to see regularities and
patterns, quickly, in data. That's why the nineteenth-century British
prime minister Benjamin Disraeli rightly talked of "lies, damned lies and
statistics." I love data of all kinds. But you have to work hard when looking at a mass of statistics to resist the temptation to jump to unwarranted
conclusions or confuse correlation with causation. As the historian of
statistics Ian Hacking wrote, "We create apparatus that generates data

that confirm theories; we judge the apparatus by its ability to produce data that fit."[1]

In everyday life, our ability to make judgments quickly and from limited information was a great evolutionary advantage, and was extremely useful when faced with a threat from a rival band or saber-toothed tiger. Yet it's not so useful when dealing with complex people and situations.

Everything we know is knowledge from the past, which may not apply in the future—the problem repeatedly stumbled on by models, algorithms, economic theories, and geopolitical dispositions, which made sense in one era, but then become dysfunctional in another. As social science has repeatedly discovered, the more you use a model, the less likely you are to question it. What starts as a pragmatic tool to answer a question becomes a truth in itself.

And so the models we use to think can also become traps. A model is held on to because it provides meaning and reassurance. Police forces notoriously cling to evidence they collect early in a case in the face of powerful contrary evidence that emerges later, and thus extraordinary miscarriages of justice result. The middle-aged cling to the theories they learned as undergraduates. Organizations become attached to models that become comfortable through use. I remember once meeting the planning team of a government that admitted that their forecasts had been no better than random guesses, but argued that the detailed forecasts were still needed to help with planning. The model became a comfort, even though it had no real use.

Familiarity also breeds blindness. In a classic experiment by psychologist Trafton Drew, radiologists examined CT scans to look for abnormalities. The indicators are tiny, and a trained eye is needed to spot them. Drew had included in the pictures an image of a gorilla nearly fifty times larger than the typical nodule. Only 17 percent of the radiologists spotted it. We see what we expect to see and what we're looking out for.

Expertise can equally entrap. As Philip Tetlock showed in his classic work on predictions, the most expert people can be the least successful at predicting the future, mainly because they become too confident in their own ability and so seek out confirming information. The best predictors are good at listening to new information, careful not to be guided by grand theories, and thus able to adjust in light of new facts.

So what follows? The implication, as the Buddha pointed out, is that intelligence has to be at war with and suspicious of itself to be truly intelligent. This was perhaps why one strand of ancient Greek philosophy believed that thinking had to be separated from practice—"thinking is out of place in action," in the words of Hannah Arendt. Deep thought has to struggle with common sense and withdraw from the everyday world of appearances. To see the truth of things, their essence, requires detachment.

To understand how collective intelligence can be sabotaged, a good starting point is a manual that was designed to do just that. In 1944, the US Office of Strategic Services (OSS, which later became the Central Intelligence Agency) published a guide for its spies in Europe on how to undermine the then-dominant German armies of occupation. Much of the spies' work involved putting bombs on railway tracks or leading Flying Fortress bombers to the right target. But some of the work was subtler. The manual recommended eight methods for undermining organizations from the inside. "Insist on doing everything through channels. Never permit shortcuts to be taken to expedite decisions." "Make speeches. Talk as frequently as possible and at great length. Illustrate your 'points' by long anecdotes and accounts of personal experiences." "When possible, refer all matters to committees for 'further study and consideration.' Attempt to make the committees as large as possible—never less than five." "Bring up irrelevant issues as frequently as possible." "Haggle over precise wordings of communications, minutes, and resolutions." "Refer back to a matter decided on at the last meeting and attempt to reopen the question of the advisability of that decision." "Advocate 'caution.' Be 'reasonable' and urge your fellow conferees to be 'reasonable' and avoid haste, which might result in embarrassments or difficulties later on." "Be worried about the propriety of any decision. Raise the question of whether [it] lies within the jurisdiction of the group or whether it might conflict with the policy of some higher echelon."

The most common OSS-type response in contemporary organizational life is to call for more analysis, data, and scenarios, especially to guide decisions where there simply aren't enough hard data to tell you what to do (guaranteeing a big diversion of time and energy from actually achieving anything). Another is to oscillate between asking for shorter papers that synthesize the key issues and then complaining that there isn't enough detail to make a decision. Again, this can be guaranteed to soak up time and energy to little effect. Another good technique is to demand a general policy on issue x rather than a specific decision.

This absorbs much management time and, as policies cumulate, clogs up the organizational arteries.

All these tendencies are perfectly reasonable. On their own, each tendency makes perfect sense. When decisions are difficult, everyone can agree on these diversionary options. But taken together, they lead to stagnation and crush any hope of creativity.

The OSS prescriptions are recognizable features of behavior in any large organization and make sense because they are not obviously disruptive.[2] Each goes with the grain of what appears to be intelligent behavior. Taken together, though, they guarantee that collective intelligence seizes up, choking on itself.

That's why the mark of good leadership—whether within management teams or on committees and boards—is that it faces these tendencies down, sometimes ruthlessly, and forces people first to make decisions and then to act. Much of the time the difficult questions are then better answered through action than they ever could be through talk and analysis alone.

An Ethic of Doubt

The Russian proverb *doveryai no proveryai* (trust, but verify) was adopted by Ronald Reagan during his negotiations with Mikhail Gorbachev. It's a useful starting point for any kind of collaboration in a network or with strangers. Start off with a disposition to cooperate and be open. But at each stage check and verify.

Any group is defined by what it keeps out as much as by what it is. It has to exclude, reject, and ignore. It can only focus on a small proportion of the data available to it and only a small proportion of the inputs that could come from the environment that surrounds it. Otherwise it would be overwhelmed.

To survive it has to select, and the principle of "trust, but verify" helps in selecting both who to cooperate with and what to believe. Groups need to go further than this, though. Like an immune system, they have to be able to push out threats—direct challenges, contrary truths, and even perhaps attitudes that are corrosive, like cynicism. To survive, a collective has to recognize the things that may bring it death or disease, and fight them preemptively.

This requires what I call self-suspicion. As human civilization has learned more about its own consciousness, it has also learned much about the tendency of individual brains, organizations, and cultures to deceive themselves. We tend to exaggerate the coherence or meaning of the world. We cannot help but select information that is convenient or comfortable. Indeed, the more we know, the more we reinforce rather than challenge our own knowledge.[3]

We remain attached to narratives and explanations long after they have ceased to be useful. These patterns, with their various names such as *confirmation bias*, *groupthink*, and *investment effects*, are visible every day in any organization as well as ourselves and our friends. They take extreme form in dictatorships or the mind-set of figures like President Donald Trump's former press secretary, Sean Spicer, who when challenged on an obvious untruth, said that "sometimes we can disagree with the facts."[4]

But these are extreme forms of everyday patterns. One of the many painful discoveries of the modern world is that beauty and truth are not identical (as John Keats, the poet, promised), or even natural cousins. What is beautiful can be a lie or truth, and a truth can be mundane or ugly.

The scientist Michael Faraday wrote particularly well on this: "We are all, more or less, active promoters of error. In place of practicing wholesome self-abnegation, we ever make the wish the father to the thought: we receive as friendly that which agrees with [us], we resist with dislike that which opposes us; whereas the very reverse is required by every dictate of common sense." And so he recommended mental discipline—"that point of self-education which consists in teaching the mind to resist its desires and inclinations, until they are proved to be right."[5]

Civilizations also develop strategies of self-distrust as well as conformism. Every society reproduces itself in terms of thoughts and institutions through what the sociologist Pierre Bourdieu called *doxa*: "an adherence to relations of order which, because they structure inseparably both the real world and the thought world, are accepted as self-evident."[6] But the contrary traditions asked questions—perhaps going back to Socrates and beyond—so as to undercut what appears self-evident. This is actually one of the definitions of wisdom: "To be wise is not to know particular facts but to know without excessive confidence or excessive cautiousness. . . . [W]isdom is an attitude taken by persons towards the beliefs, values, knowledge, information, abilities and skills that are held, a tendency to doubt that these are necessarily true or valid and to doubt that they are an exhaustive set of those things that could be known."[7] This can be a chilly virtue—the description Isaiah Berlin used of liberalism. Yet it is an essential one too.[8]

As we mature, we learn to question our assumptions—the models of the world that we've inherited from our family or social class—even if we then choose to readopt them. Peeling away, deconstructing, and reconstructing is what we learn to do as we cultivate intelligence, and the mark of a healthily intelligent society is that enough people can see its apparently natural laws and institutions as inventions that can be challenged as well as changed.

Here I want to focus on how to embed this kind of self-suspicion—the doubt that comes before wisdom.[9] One of the most consistent strategies for anchoring truth is to find solid points that are furthest from consciousness and thus from the risks of deception and self-deception. These are the physical facts of the world and the facts of number. These help to ground knowledge in ways that are resistant to imagination (though we soon learn that numbers are less objective, and more shaped by context and power, than they seem at first).

So we see mathematics as a purer truth than any other and especially the mathematics that is corroborated by the natural world. Best of all are the abstract predictions that appear surprising and are later confirmed by some measurement of the stars or atoms. We admire the practical skill of the engineer and the architect because, however glossy their brochures and eloquent their speeches, what ultimately matters is whether their bridges or buildings stand up.

Number too becomes a protection against deception, which is why the modern world loves statistics, data, and also the newer ways of seeing the world that are direct rather than mediated by representations (like the Dove satellites I mentioned earlier that directly measure economic activity on the ground).[10] The natural sciences came to view the observation of big numbers, seen from far away, as superior to knowledge of things close at hand. This—the world of cosmology, physics, chemistry—became synonymous with science. Numbers helped to generalize—to generate universal laws about objects that are assumed to be similar.

But numbers can also be used to understand the local and the specific and a very different trajectory for science would have taken this idea seriously.

As Bruno Latour put it, "what is really scientific is to have enough information so as not to have to fall back upon the makeshift approximation of a structural law, distinct from what its individual components do."[11] Such could have been an alternative direction for modern science. But it would have struggled to answer the questions of suspicion. And so the formula of stripping away the contextual to find the universal prevailed.

The randomized controlled trial (RTC) has gained a high reputation precisely because it is such a powerful tool for self-suspicion. Testing a drug, or an educational program, on one group while keeping another as a control is a demanding standard and one that is very hard to game. There are countless problems with RCTs. They are ill suited to many questions and, in medicine, have often been shown to generate wrong results. They may tell you what works, but not when, for whom, or where (and many RCTs have misled because they appeared to answer these questions, but did not). But they have a special status because they have a healthy in-built suspicion.

We need these tools for suspicion because power and identity pull intelligence toward them—they are like magnetic forces of attraction, through the desires for survival and conformity and the human wish not to be alone. These all make us see the world as we want to see it, full of confirming facts, emptied of inconvenient truths.

These magnetic pulls affect everything. It's notorious that governments see in a particular way. They overturn the lived experience of spaces in favor of a distant view, or the view of someone needing to manage and manipulate. Just reflect on how differently you think about your children or friends from the way a retailer or the government uses data to understand them. The first is rich, contextual, dense with meanings, the second is simple, standardized, and unreal.

The corporate gaze is flattening too: it shows consumers, segments and people defined by what they have bought. Whenever a real person is shown the profile built up by a social media company or market research they are shocked; it's like a strange cartoon, with some recognizable features but nothing like the identity and stories that really define what the individual is. Data is particularly flattening, because it captures actions and occasionally stated preferences, but almost never meanings and contexts. And official data by its nature knocks off the fuzzy edges and can stand at odds with how people define themselves, for example, in census definitions of ethnicity.[12]

So doubt is a crucial tool for shared intelligence. But it has to be doubt with limits—doubt that can be suspended for action, doubt that wisdom learns sometimes to put aside, because doubt can also subvert intelligence. Here are the words of a tobacco industry executive in the late 1960s, nearly two decades after incontrovertible proof had shown that smoking caused lung cancer: "Doubt is our product since it is the best means of competing with the 'body of fact' that exists in the minds of the general public." More recently, a similar strategy was followed by the oil industry to confuse and disorient the public on climate change, focused on the ambiguities and unexplained details to be found in any field of science to confound a bigger truth.[13]

FIGHTING THE ENEMIES OF COLLECTIVE INTELLIGENCE

Intelligence is highly improbable, and collective intelligence is even more so. It runs into misinformation, misjudgment, and misunderstanding. This is unavoidable because thinking is a site for conflict, tactics, and strategies. Insight jumps out of the clash of argument as well as linear discovery of truths. The happy accounts in which collective intelligence emerges naturally from large groups of people sharing honest information miss at least half of the truth.

Every real group combines multiple identities, interests, and wills, and is a container for competition, games, and feints. This becomes apparent in such mundane examples as the struggle to edit a Wikipedia page about a politician, attempts to manipulate data or corporate profit figures, and battles over reputation.

The enemies of collective intelligence include deliberate distortion and lies, noise, misobservation and misinformation, trolling, spamming, deliberate distraction, fear of the unknown, prejudice and bias, and "overstanding" taking the place of understanding.[14] They are now joined by threats from machines—like artificial intelligence botnets that have brought communications systems to a standstill or viruses.

These are part of the daily currency of human (and machine) interaction, amplified in digital environments. Today's social networks need protections against cyberattacks, viruses, and worms to function at all, fighting against trolling—targeting an individual with hateful communication, tweeters and bloggers who value provocation over truth, tricksters planting false stories and riding the waves of gossip as well as eager sharing, criminals spreading "ransomware" or hackers launching "denial of service" attacks, flooding servers with information or demands, all protected by anonymity. As Peter Steiner wrote in the *New Yorker* in 1993, on the Internet no one knows you're a dog, and so behind anonymity, anything goes and anything can go.

In this arms race, networks have built up their own arsenal of verification tools. These tools include reputational devices, as on eBay or Uber, to separate the bad trader from the good one or bad taxi customer from the safe one. Captcha tools try to distinguish real people from machines. Others have fought against spammers, encouraging users to classify spam feeds so that they benefit from pooled judgment. This combination of human input and machine learning has so far maintained its lead in the arms race with enemies (and depends on keeping some things hidden, such as not showing antispam rankings that would make it too easy to reverse engineer a better way of overcoming the antispam barriers).[15] There are some limits: artificial intelligence still struggles to keep up with the fertility of human insults.[16]

The Ushahidi team developed algorithms using machine learning to classify such sources as Twitter accounts so as to filter out less reliable sources. There are now many tools available for analyzing whether pictures have been modified, or combining human input and machine learning to spot whether pictures have the right light or landmarks, such as in a photo or video from the Syrian Civil War that may be used by news services. The more contested the field, the more important these countermeasures are.

Then there are the identifiers that track down the computer behind denial of service attacks. The sophisticated systems for overturning anonymity struggle with the burgeoning power of the dark web and strong cryptography. Wikipedia has a range of tools to stop editing wars and toxic content; most of the networks use recommendation and reputation systems to prevent trolls. The many sites gathering feedback from customers are prone to fraud—say, restaurants and hotels posting positive views on themselves—but algorithmic tools can try to spot these frauds (Yelp's algorithmic indicator, for example, found that 16 percent of restaurant reviews are fraudulent and tend to be more extreme than other reviews). By and large, these tools work reasonably well, but their success depends heavily on how well communities build up strong social norms and cultures.

Positive network effects are familiar—the extra value that comes from a new user of Facebook or WhatsApp so that the growth of a network increases its value; cell phones and science systems are good illustrations. Functional networks obey Metcalfe's law, and the explosive growth of services like Twitter and Instagram shows just how powerful these positive network effects can be. Some social networks along with networks involving power and unique information, however, do not obey the same patterns. There may be more value in a much smaller network with a stronger commitment to truth and mutual support. Negative network effects are found when increasing the size of the network diminishes its value or effectiveness.

We know that there is evidence of strong positive network effects for crowdsourcing data, analytic capability, or memory, but also some evidence of negative network effects for intensive deliberation and judgment (and evidence that large groups may take riskier decisions). These partly result from the negative impact of larger networks on attention—floods of Facebook requests or irrelevant data get in the way of valuable activity—and partly from a weakening of responsibility. They can worsen other well-known vices of digital tools: the tendency to promote binary choices, simplistic categorizations, or anonymity encouraging aggressive behaviors. New norms as well as technologies may be essential, such as for restricting anonymity.

We shouldn't be surprised that much more powerful tools for linking people together do not automatically produce greater mutual awareness and understanding. One of the more surprising findings of linguistics is that population density roughly correlates with language diversity—almost the opposite of what you might expect. We use language, in other words, both to communicate and out-communicate, and we use networks to do the same.

PART III

Collective Intelligence in Everyday Life

HAVING SET OUT THE CONCEPTS and theories we need to understand collective intelligence, the next few chapters look at familiar fields to show how we can make sense of them and prescribe better ways of thinking.

Almost every aspect of daily life is in some way shaped by the intelligence of the systems around us. This includes an economy that provides work and goods; democratic institutions and governments that represent, support, and protect us; universities that help us understand; and the millions of meetings that take place every day to help us cooperate with others.

Each chapter is a sketch, a pointer to what I hope will in time become a much richer field, supported by empirical evidence, which can address how well different fields observe, reason, create, and remember, and guide them to do better. Alfred Binet invented the IQ test as a tool for diagnosis that could help people enhance their intelligence. We have nothing comparable for collective intelligence. But my hope is that through combining theory and practice, we can move before long toward much more effective diagnosis and prescription.

- 11 -

Mind-Enhancing Meetings and Environments

> I've searched all the parks in all the cities and found no statues
> of committees.
> > —*G. K. Chesterton, Trust or Consequences*

MANY OF US SPEND MUCH OF OUR TIME in meetings. They are the everyday expression of collective intelligence—bringing groups together to think. But often they feel like a waste of time, and fail to make the most of the knowledge and experience of the people present. Oddly, the vast majority of meetings in business, academia, and politics ignore almost everything that is known about what makes meetings work.

In this chapter, I look at what is known, and how that knowledge can be used. I explain why meetings haven't disappeared, despite an explosion of technologies that might have rendered them redundant. And I suggest how meetings might be organized to make the most of the collective intelligence in the room and beyond, before turning to how physical environments can be shaped to amplify individual and shared thought.

THE PROBLEM WITH MEETINGS

The formats used for meetings are old. Most organizations still depend on the board or committee, usually made up of between five and twenty people, for the most crucial decisions. This remains the supreme decision-making body in organizations as varied as Ford and the Politburo, Greenpeace and Google (with twelve sometimes treated as the ideal number). At the level of the nation, we still depend on parliaments and assemblies, usually made up of a few hundred individuals, which meet in formats often little changed over centuries. For more everyday matters, there are committees, teams, or workplace meetings. For the world's major religions, there are the often-rigid formats of service, prayer, and song. And for the worlds of knowledge and ideas, there are conferences and seminars, with anything from a few dozen to a few thousand participants—again, formats similar to their equivalents a century or more ago.

We depend greatly on these old forms of face-to-face deliberation, and the advance of technologies for communication across space and time has done little to displace them.

But our very dependence fuels frustration. The typical meeting barely attempts to make the most of the knowledge and experience in the room. The loudest or most powerful speak the most, drowning out the weak or shy, and much that should be said isn't.

Many have tried to develop more open, lively alternatives. There are boardrooms like the one at Procter and Gamble that are surrounded by screens, where the entire global leadership team meets weekly (physically and virtually) to review data on sales, margins, or customer preferences. There are cabinet rooms like the one used by the Estonian government, with screens instead of paper. Some companies have gone to extremes to minimize the curse of meetings. Yahoo! sets ten or fifteen minutes as the default for meetings. Others hold meetings standing up. To counter the torrents of useless talk, some cultivate silence. Amazon requires six-page memos to be prepared before any meeting and then read in silence by each person for thirty minutes before a discussion. Some conferences have experimented with giving participants buttons that they can press when they want the speaker to stop—a wonderfully empowering idea, but sadly far from widespread.

Another cluster of innovations has tried to reduce the need for people to congregate together physically. Telepresence meetings and Google Hangouts, online jams involving thousands, smaller webinars, and meeting tools like Slack allow teams to meet and work online.[1]

Yet another set of innovations turn meetings inside out, making the formal parts of meetings more like the informal conversations on the sidelines that are frequently so much more enjoyable and memorable. This was the prompt for "open space" methods several decades ago as well as unconferences, World Cafés, Flipped Learning Conferences, Holocracy, and other tools for democratizing larger gatherings, all designed to overturn the stiff formality of the traditional meeting so that anyone can propose topics for discussion and participants can choose which conversations to take part in.

Some echo the unstructured worship and creative use of silence pioneered by the Quakers four centuries ago.

Such tools can be a refreshing alternative to the stultified, over-programmed conference formats of keynotes and panels. But they can also be frustratingly vague, making the whole less than the sum of the parts; they can be hard work to organize, too, and aren't well suited to sustained problem solving. An odd feature of these innovations is that they tend to crystallize quickly into a formula—and don't then evolve in response to experience. Oddly, too, they worsen some of the tendencies of bad meetings, such as dominance by extroverts.

Hybrids seem to work better, like the meetings held by the Dutch social care organization Buurtzorg that combine a flat structure with fairly strict rules to ensure that decisions are taken. To adopt the ideas of culture theory, the best meeting models combine elements of hierarchy, egalitarianism, and individualism; the ones that are either too purely hierarchical or too purely egalitarian (like open space) work less well.[2]

WHY SO MANY MEETINGS?

The British Civil Service was not keen on meetings, and until recent decades many of its office buildings had no meeting rooms. It preferred minuted reports—sent from desk to desk. By contrast, a recent study found that on average, 15 percent of an organization's collective time today is spent in meetings, with senior executives spending two days a week in meetings with three or more coworkers.

This spread can be seen as a horrible creep of unproductive time. But it's better understood as a logical response to the growing complexity of today's decision-making needs. When power relationships are ambiguous, problems are complex, and the environment within which decisions are being made is itself changing rapidly, we benefit from regularly coming together to realign goals, interests, and attitudes. This happens most easily through conversation, and is harder when decision makers can't see each other's social cues. Misunderstandings are more common over e-mail than phone and over the phone than with video messaging. There's also strong evidence that we're much happier interacting with others face to face than virtually.

Even the most banal procedural meetings help participants to gauge one another's interests, attitudes, and relationships. That helps them negotiate more easily as well as develop a shared intelligence and culture, though it can also help a hierarchy to enforce conformism and squeeze out deviance. The same is true of activities that cut across many organizations—working in formal partnerships, alliances, supply chains, networks, and joint ventures. These too require meetings (as well as a mushrooming quantity of e-mails and conversations) to coordinate actions. Only a small proportion of issues can be handled through formal contracts. Meetings cost a lot (and there are now devices to work out exactly what that cost is).[3] But not holding meetings can be even costlier.

There is a vast research literature analyzing precisely how meetings do or don't work, the subtle strategies we use when talking to others, and the role played by supporting activities, such as providing agendas, documents, minutes, presentations, preparatory e-mails, and exchanges.[4] To work well, they have to counter our tendencies. One is our desire for social harmony, which means that in teams, people tend not to share novel or discomforting information. Another is that our egos tend to become attached to ideas and proposals, and so make it harder for us to see their flaws. A third is our tendency to defer to authority. And a final, opposite one is that although we all make judgments about whose views we respect, and recognize that in any group the value of contributions will vary greatly, we often default to equality, giving equal weight to everyone—an admirable democratic tendency that unfortunately can mean that poor quality contributions crowd out better ones.[5] All these tendencies, if unchecked, lead to worse decisions and degrade collective intelligence.

So what could make meetings better? To someone with a hammer, every problem looks like a nail, and in the same spirit most organizations get stuck in habits, using the same meeting formats regardless of what they're trying to achieve. But there are usually many other options worth considering. Here I summarize some of the crucial factors that can help to make a group more like a collective intelligence, and less like a miserably boring committee or conference.

Visible Ends and Means

A first step is to ensure that the purposes, structures, and content of the meeting are well understood by all participants. Is the meeting to share information, create something new, pray, or make a decision? It will need a different shape depending on which of these is the main goal.

Agendas that are easily accessible beforehand ensure valuable time is not wasted and everyone is up to speed the minute they walk in (and agenda setting can either be done by the most senior person or in a more open way). Sharing background papers and materials encourages a common understanding of the purpose of the meeting, and many digital tools can make these visible.[6] This doesn't imply that all meetings should be instrumental. Some meetings should be open-ended and exploratory. The point is that this should be clear.

ACTIVE FACILITATION AND ORCHESTRATION

Even the most motivated groups don't self-organize themselves well. That's why the role of the chair or facilitator is so important for getting good results. The role doesn't have to be filled by the most powerful person in the room; it may be better played by someone junior, given the temporary authority to ensure the meeting achieves its purposes and sticks to time.

To do their job well, they need to keep the meeting focused on its goals. Yet they can also help the group to think well by countering the risks of anchoring (the first person to speak sets the agenda and frames). They can work hard to avoid the risks of unequal contribution (with higher status counting for more than greater knowledge and experience). Other methods like leaving periods of silence for groups to reflect and digest can improve the quality of discussion. So can encouraging participants to write down their most important thoughts before the meeting, allowing the most junior person to speak first (as in the past in the US Supreme Court) or interweaving different scales of conversation (from plenary to smaller groups, down to discussions in pairs and back).

EXPLICIT ARGUMENT

Good meetings encourage the explicit articulation and interrogation of arguments, and ideally allocate people roles to interrogate them. These roles can be formalized or left more informal. The key is to avoid skating over the uncomfortable aspects of disagreement.

Psychologists have shown that people have a strong *confirmation bias*. This means that when we reason, we try to find arguments that support our own idea. At an individual level, this can lead us to make bad decisions. But from a collective point of view it can be extremely effective, since it encourages people to develop the best versions of their arguments. Confirmation biases cancel each other out and then push the group to a better solution.[7]

Giving a structure to argument then becomes an important design challenge. Parliaments do this through formal debates, and courts through the presentation of evidence and interrogation of witnesses. Some hedge funds interestingly incentivized disagreement, rewarding the people who had disagreed with trades that then turned out to succeed (and the account by Ray Dalio, founder of the world's largest hedge fund, Bridgewater, is a good summary of the value of encouraging argument and criticism).[8]

For argument to work well, meetings benefit from structured sequences—so that, for example, discussion focuses first on facts and diagnosis, before moving on to prescription and options (which tend to be more fraught as well as more bound up with interests and egos). The general point is that conscious, deliberate processes usually improve the quality of discussion.

MULTIPLATFORM AND MULTIMEDIA

The best meetings use multiple tools in parallel. They combine talk and visualization, and small talk as well as plenaries involving everyone. A consistent finding of much research is that people learn and think better when supported by more than one type of communication. Information presented in different forms aids learning and understanding. A written five-page report, presentation, and selection of images combined with a verbal discussion will have differing effects, but can add up to a better understanding of the issues. This is also why simple rules can help, such as no numbers without a story, and no story without numbers, or no facts without a model, and no model without facts.

Digital tools help visualize complex ideas at meetings, making it easier to reach decisions, such as allowing ideas to be visually collated by multiple people in real time, creating linked networks of ideas collaboratively, or presenting complex data in more accessible ways.[9] Other promising tools show the participants how ideas and arguments are evolving, and how the group mind is thinking and feeling.

REINING IN THE EXTROVERTS, OPINIONATED, AND POWERFUL

Social psychologists using survey and observational techniques to measure group intelligence have shown that they correlate only partly with the average and maximum intelligence of individual group members. For example, one recent psychology study found that three factors were significantly correlated with the collective intelligence of a group: the average social perceptiveness of the group members (using a test designed to measure autism that involves judging feelings from photographs of people's eyes), relatively equal turn taking in conversation, and the percentage of women in a group (which partly reflects their greater social perceptiveness).[10]

Extroverts dominate the typical meeting. As a result, many participants may not feel comfortable contributing. Formats that make it easy for everyone to contribute, rein in the most vocal, and give people time to think before speaking are likely to work better.

PHYSICAL ENVIRONMENTS THAT HEIGHTEN ATTENTION

Meetings benefit from environmental conditions that make it easier to pay attention to the meeting itself and other participants. That includes sufficient natural light, quiet and space, and giving people chances to move around (and not staying seated for more than an hour or two in any one stretch).[11]

Physical shape also influences the quality of meetings. For instance, square or circular meeting spaces allow everyone eye contact with everyone else and so encourage greater engagement. The classic boardroom table is a poor design from this perspective, as is the classic theater-style conference hall.

Finally, some organizations ban use of laptops or smartphones during meetings—partly to ensure full attention. US Cabinet meetings, for example, require participants to leave their phones at the door.

DELIBERATE DIVISIONS OF LABOR

The best meetings take advantage of a division of labor with distinct roles, including facilitation, record keeping, synthesis and catalysis, court jester, and professional skeptic. They then end with explicitly distributed tasks given to participants.

For the meeting itself, methods that distribute roles among participants include Edward de Bono's "six thinking hats," where different-colored metaphoric hats open up different perspectives to thinking.[12] White addresses the facts and what is known; black provides caution and critical thinking; red emphasizes feelings, including intuition and hunches; blue manages the process, ensuring it is followed by the group correctly; green promotes creativity, new ideas, and options; and yellow encourages optimism, looking for values, benefits, and advantages. The idea is that the interaction of these viewpoints leads to better outcomes, especially when participants try out different roles rather than becoming fixed in just one.

David Kantor's "four player model" has a similar approach.[13] Groups are divided into four roles: movers, who initiate ideas and offer direction; followers, who complete what is said, help others clarify their thoughts, and support what is happening; opposers, who challenge what is being said and question its validity; and bystanders, who notice what is going on and provide perspective on what is happening, offering a set of actions people can take while in a conversation. In a healthy meeting, people will move between these roles.

It's easy to imagine other variants; what's important is to formalize differentiated roles (which many of us hear as inner voices when we're trying to make a decision). Even better, tasks are distributed to named individuals at the end of the meeting, so that these can then be tracked. Knowing that this is going to happen makes it more likely that people will pay attention.

MEETING MATHEMATICS

There is no perfect mathematical formula for meetings, but experience suggests something close to a law that correlates the complexity of the task, number of participants, available knowledge and experience, time, and degree of shared language or understanding. This is particularly true for meetings that aim to come to a conclusion or make a decision.

The most common reason meetings fail is that they don't conform with meeting mathematics: there are too many people or too little time, too little relevant knowledge and experience, too sprawling a topic, or insufficient common grounding.

A simple task, with few participants, and well-understood common language and references, may lead to quick results. Whereas a complex task, with many participants and not much shared frame of reference, may take infinite time to resolve, and even if the time isn't infinite, it may feel so.[14]

In framing understanding of an issue or mapping out options, diversity brings great advantages, as does tapping into many minds. But translating that diversity into good decisions usually requires the added element of a common grounding or culture. So strong organizations try to bring in a diverse workforce as well as tap the brainpower of their partners and customers, and then funnel decisions through a group that also has a strong common understanding and language along with a depth of relevant knowledge. On their own, crowds aren't wise.

Getting the mathematics of meetings right is key not just for face-to-face meetings but also for online ones. In principle, online meetings can gather in much more intelligence—knowledge, ideas, observations, and options. The most successful online collective intelligence projects tend to combine quite precise tasks and reasonable amounts of time, and are more about gathering and assembling than judging. As a result, they don't require so much common framing or the subtle cues needed for ongoing collaborative projects.

Good Meetings Are Visibly Cumulative

Meetings rarely happen in isolation. Some two million hours of work may go into the design of a large building or car, including many hundreds of meetings. There are complex tools to coordinate the efforts of a large work team as well as simple devices like feedback forms and regular reviews that link any meeting to previous ones on the same topic (through traditional means like minutes, or more modern ones like data dashboards and lessons learned exercises). Social media patterns can be analyzed to show

how people interact after meetings, and social network analysis tools can be used to reveal underlying patterns of helpfulness in organizations and across them (for example, surveying who people rely on to get information or get things done).[15]

Within workplaces, evidence suggests that the quality of relationships and attitudes, measured by the quality of small talk before meetings, matters more for meeting effectiveness than good procedures on their own.[16] Participants' perceptions of meeting effectiveness have a "strong, direct relationship with job attitudes and wellbeing."[17] If attendees of a meeting are happy going in, they will be productive throughout, and happier and more productive afterward. Modest tools can influence this.[18] So can simple devices like encouraging everyone to take coffee or lunch breaks at the same time.[19]

Avoidance

Newspapers and news shows fill up their space regardless of how much news has happened. The same is often true of meetings. Organizations schedule regular cycles of committee, board, and group meetings, and then feel impelled to fill up the available time. This is one of the sources of frustration and boredom in many organizations, because it means that many meetings feel pointless. An alternative is to leave time slots in, but more frequently cancel meetings when they're not needed, radically shorten them to align with the number and seriousness of issues needing to be addressed, and consult with participants on whether the meeting is needed, and if so, how long it should be.

Often people feel uncomfortable canceling meetings for fear that it implies that no work is being done. Similarly, people in big bureaucracies feel uncomfortable not attending meetings—for fear that they may miss out on vital decisions or not be seen as a team player. The opposite is a better approach—canceling or shortening meetings as a sign of effective day-to-day communication

What makes meetings effective mirrors the framework set out earlier. For a meeting to be most collectively intelligent, all five of the approaches to organizing collective intelligence need to be in play. Autonomy: the meeting may take place within a highly structured hierarchy, but it will achieve the best results if it is allowed some autonomy—the freedom to explore and open up before options are closed down and decisions are taken. Balance: the meeting needs the right mix of types of intelligence depending on its task; it may be primarily to gather observations, create, remember (for example, synthesizing lessons learned), or judge, but will benefit from clarity on what it is trying to achieve and how. Focus: a clear goal for the meeting makes it easier to determine which contributions are relevant or not, and avoid sidetracks. Reflexiveness: taking stock of whether the meeting is on track and new categories of thinking methods are needed. Integration for action: the role of the chair or facilitator to integrate contributions, and then turn complex, flowing conversations into the beautiful simplicity of a good decision that can be acted on.[20]

Although this is far from being a science, much is known about what makes meetings work. Yet strangely, the great majority of meetings ignore what's known, wasting billions of hours of precious time.

Digital technologies that help us design and manage meetings may encourage more hunger to use this knowledge. For now, they offer only modest ways to enhance meetings: making them easier to organize, bringing disparate groups together, and more recently, showing how the group's conversations are evolving. In the not too distant future, however, we may use computer facilitators much more to regulate time, ensure everyone has a chance to speak, suggest or manage strategies to overcome impasses, monitor emotions through scanning faces, and help avert unhealthy conflicts.[21] They may also be good at coaching people how to handle difficult meetings, though we may end up coming to prefer these roles to be played by inhuman machines rather than self-interested leaders.[22]

ENVIRONMENTS THAT SERVE INTELLIGENCE

So the formats of meetings can either enhance or diminish the brainpower of participants. But can the same be said of places and buildings? How can they make people collectively smarter? Over many decades, Jim Flynn showed that intelligence was slowly increasing around the world. His work was once controversial, but is now largely accepted, though the pace of increase appears to have stalled. It immediately raises the question, Why? The answers are fuzzy, yet it's clear that some types of environment tend to enhance intelligence. Flynn's work suggested that immersion in cultures full of abstraction, metaphor, and simulation left people smarter as well as better able to reason conceptually. Environments that are stimulating, demanding, and engaging will tend to enhance intelligence, while those that are full of fear and unresponsive, or conversely too predictable, may do the opposite.[23]

The idea of consciously shaping a mind-enhancing environment is not far-fetched. To guide this idea, we can look to what is known about environments specifically designed to aid thought. Intelligence happens in any environment. But it is assisted by the quiet of a cell or study, the concentration of the seminar room, the light of the quadrangle, and the tools at hand in the laboratory. There is a fair amount of research on why some patterns appear common in places devoted to thought, such as why high ceilings along with a sense of abundant physical space are so often chosen for libraries or universities, perhaps because they encourage expansive thinking. Walking also seems to play a role, which is why quadrangles are so popular in universities, and why cities encourage parks, squares, and gardens that combine gentle stimulation, variety, and an excuse for movement.

The best environments mix open spaces that encourage serendipitous interaction and the quiet, private corners needed for an intense conversation or meditation. The fashion for wholly open plan offices is at odds with what's known, and can be just as much the enemy of thought as the traditional overcompartmentalized building that makes it hard to have conversations in corridors or doorways.

At one extreme, intelligence can be literally embedded into environments. Mark Weiser, the founder of ubiquitous computing, suggested that "the most profound technologies are those that disappear. They weave themselves into the fabric of everyday life until they are indistinguishable from it."[24] We already have a generation of buildings able to read their occupants, show who is moving where and which people interact with each other, or reveal what patterns of external communication are taking place. This is the city of the Internet of things, frictionless, constantly generating data about every activity, and potentially feeding that information back to help people think about their thoughts and actions. Virtual and augmented realities already have the potential to accelerate learning as well as accentuate brain plasticity—for example, by letting you directly copy and perfect the tennis strokes of a master. It's intriguing to wonder what might happen if these could be embedded into physical environments, such as parks designed for stimulus that mixed the virtual and physical worlds.

These may work best by being difficult. The risk is to make things too easy and dumb down through rendering motion effortless. The effect is similar to that of play areas for children that were progressively redesigned to reduce the risks of any injury, with rubber surfaces in place of stone and tarmac along with equipment that avoided anything too angular and jagged. Yet the result was that children were less prepared for the world they had to live in and less resilient in terms of the inevitable frictions of growing up. And so the ethic of design changed again, to carefully calibrated risk with stones and uneven surfaces.

There's a similar evolution for adults. The aim of twentieth-century city planners was to make the city frictionless and promote ease of motion in traffic, or for pedestrians, to reduce the mental clutter of the city so as to make passage through it as mindless as possible. But an opposite ethos is also emerging. It involves deliberately reintroducing friction and stimulus, adding prompts that are not just the mental clutter of mass advertising, and allowing spaces for silence and tranquility, free of communicational pollution, but also sometimes injecting surprise or challenge through walls that talk and images that arrest.

Within the design of buildings, a lot has been learned about how physical design shapes or at least influences thought. Thomas Allen discovered at the Massachusetts Institute of Technology in the 1970s what became known as the *Allen curve*: an exponential relationship between distance (how far two engineers sat from each other), and how regularly people communicated with each other. Someone sitting six feet away from you was four times more likely to talk to you regularly than someone sitting sixty-five feet away. People working on different floors were unlikely ever to interact, and people in separate buildings would probably never be encountered at all. Of course, many devices could be used to overcome these boundaries, such as joint exercises, social events, or deliberate pairings. Yet the main message was that proximity influenced both the quantity and quality of interactions, and made it more likely that people would have the kinds of informal interactions that generate more random, creative ideas.[25]

A different perspective comes from research on color, such as the suggestion that exposure to 480 nanometers—a vivid blue—before 11:00 a.m. stimulates attention. We know that round buildings are disorienting (people literally get lost), and spaces can be made more useful with writable surfaces and plenty of screens. But for now, despite intriguing hints, there is little evidence on mind-enhancing environments, though the study of play and games offers some pointers. We have revealed preferences—for example, for white doors inside homes, and dark or black ones outside (but not the reverse); natural light for daytime activities; and lines of sight, but also lines of escape. There is little hard evidence, however.

As understanding of mind-enhancing environments evolves, we might expect the discourse of the smart city to change. It has focused almost exclusively in recent years on hardware (sensors tracking flows of cars, energy, or water, say). But what if the question turned to how cities could make their citizens smarter? What if cities reined in some of the potential enemies of smartness (which range from unwanted communications that clutter the mind to drab, soulless spaces and fear)? And what if they worked harder to enhance stimuli, and make the experience of living in the city more stretching and enlivening? Here lies another route for the evolution of collective intelligence.

- 12 -

Problem Solving

How Cities and Governments Think

PERHAPS THE WORLD'S GREATEST SUCCESS of collective intelligence in the twentieth century was the elimination of smallpox—a rare illustration of a solution as big as the problem it aimed to solve.

This project to eliminate a disease that had killed millions was subtle and comprehensive. It combined observation, highly decentralized interpretation, creative design of new vaccines, and constant feedback and iteration, including orchestration of the shared memory of what worked.

Much of the credit went to Donald Henderson, the official put in charge of designing and running the program. But some of the credit should also go to a forgotten individual, Viktor Zhdanov, the deputy minister of health for the Soviet Union. In 1958, he attended the Eleventh World Health Assembly meeting in Minneapolis, after the USSR had boycotted the organization for nine years. Zhdanov set out a long, detailed, and visionary proposal to eradicate smallpox within a decade, which promised to be the first time any disease had been eradicated completely. This was a project without precedent—but it was conveyed with passion and optimism. Zhdanov showed that it was solvable partly because it was a disease borne by humans rather than mosquitoes. He also demonstrated how well the USSR had already done, and quoted Thomas Jefferson's letter to Edward Jenner, inventor of the smallpox vaccine: "Medicine has never before produced any single improvement of such utility." Encouraged by the USSR's offer of twenty-five million doses of the vaccine and logistical support to poor countries, the World Health Organization changed its position and agreed to eradicate the disease.

This was an example of radical change coming from the top—a holder of power naming a problem, offering a plan to solve it, and supplying the resources needed to carry that plan through. Yet its execution depended on something more like the 360-degree collective intelligence I've described elsewhere in the book—a strong center, but also a highly distributed system for gathering intelligence and adapting the strategy in light of experience.

Some of the other greatest changes to human collective intelligence have, likewise, started both from the bottom and top—for instance, the advance of both human and disability rights, or learning to see the world as an ecology. Each of these was an example of radical, systemic thinking that involved both naming and solving problems. In the case of smallpox, the World Health Organization had to think in multiple dimensions—medical, economic, social, and political. It also had to create a rough-and-ready collective intelligence assembly—gathering data, interpreting, analyzing, innovating, and learning quickly from action on the ground.

The premise of this book is that the world could have many more transformative acts of this kind, which in retrospect look like the results of a truly global intelligence: aware of needs, aware of what has to be done, with the power to mobilize money and brains, and capable of linking thought and action.

In this chapter, I look in more detail at how complex problems—like smallpox—are solved, particularly on the scale of cities and towns, but increasingly too at the level of the world as a whole.

How Can a City Think Better?

Many cities have invented new ways of thinking about their own condition and prospects, including third-loop learning. In the nineteenth century, pioneers in statistics found new ways to map diseases like cholera or the house-to-house incidence of poverty, transforming how cities saw themselves. Later, the planning profession transformed both the macro-shaping of cities (the view from fifty thousand feet of arterial roads and ribbon suburbs) and the micro counterparts, with parks and street furniture. Cities also learned how to think together in larger groups. For example, Bilbao Metropoli-30, a multistakeholder partnership formed in the 1980s and grounded in the shared identity of the Basque region, helped transform the city's situation. There are many other examples of this kind of large-scale collaboration, often extending over many years (and there are also many cases of city stupidity—vanity, sloth, and corruption, with the Sochi Olympics in Russia a near-perfect instance of waste, with over $50 billion committed to building an extraordinary collection of underused assets, surrounded by poverty).

Visiting many city governments has confirmed for me that their tools for thought and decision making lag far behind their tools for such things as transport management or infrastructure. Too many depend on the mayor being a genius of understanding and problem solving. A better ideal is a city that can mobilize many different kinds of intelligence to help with problem solving, policy, and action.

A few years ago, I helped shape one attempt to do this. It was both a success—in that it generated momentum and practical ways of organizing a more intelligent city—and a failure—in that a change of mayor and stark public spending crisis led to it being stopped just as it was beginning to work.

The London Collaborative brought together three tiers of government: the national government, which controlled most public spending in the city; the mayor and Greater London Assembly; and the thirty-two boroughs, which were responsible for many services such as education and care for the elderly. The London Collaborative was funded by all of them, but run by a group of NGOs at arm's length from formal power.[1] The idea was to encourage more effective common problem solving across the city. It also aimed to create a sense of community by bringing a thousand leading public officials and politicians along with civil society and business leaders together in events that looked at big challenges and emerging ideas. Working groups were set up cutting across all tiers, and focused on problem solving and innovation, specifically on topics like workless households, retrofitting old homes to cut carbon emissions, and behavior change. Younger officials staffed these working groups and pitched ideas to groups of chief executives. Future-oriented scans helped to create a common sense of what was needed and possible. A rough-and-ready website made it easier for cross-cutting groups to come together to share information and ideas.

Other elements that were planned included open-data stores along with wikis for the main public agencies to share information and knowledge, such as on economic conditions in parts of the city, gangs, and transport

(Intellipedia was one of the models for this—a great innovation that, unfortunately, was tested in US intelligence agencies, just about the most unsuitable territory imaginable). A much more systematic clearinghouse was set up for commissions from academics in the main universities, including regular sessions linking decision makers and researchers on topics such as public health or gang violence. There was then intended to be a rolling process of strategy development for crosscutting issues, ideally with mutually transparent plans and shared targets across the different tiers and agencies.

Despite support among many of the key chief executives, the London Collaborative challenged the power of some existing bodies and fairly soon, a newly elected mayor killed the project off.[2]

At the time, I asked many people involved in city governance around the world what parallel solutions there were to the challenge of how to help cities to think and make the most of their intelligence. Many had research institutes, strong relationships with groups of universities, and the beginnings of sophisticated open-data policies. But I couldn't find anything resembling a collective intelligence system, and most were bedeviled by tensions between the tiers of government.

How Could a Collaborative Work?

If we were starting again, how might we design a mechanism or assembly to help a city think? A useful starting point would be to apply the frameworks set out earlier and assess a city's abilities to observe the world around it.

Rio has a famous control room, a modern-day panopticon from which the city's staff can view roads, hillsides at risk of mudslides, or capture Twitter commentary from the public. Sensors gather real-time information about air quality or noise. There are statistics (usually with a lag, but increasingly becoming close to real time) and a plethora of data, from commercial information to public data. New methods such as the UN Global Pulse team's semantic analysis of Twitter show promise as ways of predicting unemployment levels or reductions in consumer spending, responding to keywords used by people in their communication with each other.[3]

Other cities have shown how other elements of intelligence can be organized well. Dozens of cities now have labs to organize creativity more systematically (from the various teams set up in New York City by Mayor

Michael Bloomberg to the Seoul Innovation Bureau under Mayor Park Won-soon). Many now have offices of data analytics or partnerships with private companies like Waze that orchestrate transport and other data into useful forms.

We can then look at the tools used for other elements of intelligence. How is memory organized, so that mistakes aren't unnecessarily repeated? How is attention focused on what's important rather than only what's urgent? How are new patterns interpreted, and how does learning quickly follow action? Doing all these things well will, I hope, in time become part of the craft skill for experts in collective intelligence.[4]

PROBLEMS, PROBLEMS, PROBLEMS

The biggest test for the city will be its ability to solve problems or at least manage conditions more effectively.

Some like to believe that nothing is possible. But many apparently intractable public problems—like smallpox—have been solved. Chlorofluorocarbons in the atmosphere, acid rain, and in some places, the problems of public health, homelessness, or high crime have all been largely solved. Some societies have solved intense conflicts, and others have jumped from stagnant poverty to dynamic growth.

The casual fatalism that claims that nothing works is hard to reconcile with the fact that global life expectancy has risen forty years in the last century, poverty has dramatically fallen to less than 10 percent of the world's population, and deaths from violence and warfare are, proportionately, the lowest ever.

Yet many societies struggle to cope with apparently intractable problems, from climate change to obesity, inequality to racism. It's not always obvious that the world's capacity to solve problems has increased.

PROBLEM SOLVING AS AN ACTIVITY

So how does anyone, or any organization, become good at solving problems? There is a large and sprawling research literature on problem solving in all fields that shows how we discover insights and avoid errors. The literature points to the importance of both first-loop learning—applying existing methods of analysis and design—and second-loop learning—generating new ways of seeing things. It shows how a variety of styles of thinking come together in difficult problem solving—logical, analytic, verbal, and visual—and how these together help people to spot surprises, coincidences, and contradictions, and then through analyzing them, find new routes to solving problems. Sometimes we think best with analogies, stories, or images, and making use of unconscious processes (complex problems are often solved when we're not thinking about them, including when we're asleep). Attacking a problem head-on with logic isn't always the most intelligent approach.

But each creative jump has to proceed in tandem with skepticism. Much of the literature on problem solving encourages people and groups to seek out disconfirming as well as confirming data, and ask rigorously about what information or knowledge they need to make a good decision. If our possible solution involves other people, we might also need to think through how others might respond to our actions—in other words, thinking dynamically and strategically.[5] To be good problem solvers, we need to be adept at all these approaches, and know how to extend, graft, combine, and translate.[6] We also need to know how to spot the telling clue or come up with entirely novel ideas.[7] For the city seeking to solve problems better, all these capacities need to be brought together into a deliberate framework that's both wide and deep.

The people who will then be best suited to creative problem solving are likely to be very different from the typical bureaucrat. A major study on the types of people responsible for complex breakthroughs reported that "individuals who have high cognitive complexity tend to be more tolerant of ambiguity, more comfortable not only with new findings but even with contradictory findings. Moreover, such individuals have a greater ability to observe the world in terms of grey rather than simply in terms of black and white . . . they report that learning new things and moving into new areas is like play . . . [and] tend to be more intuitive and have a high degree of spontaneity in their thinking."[8]

A good example I was involved in that brought together both people and methods fit for problem solving aimed to sharply reduce the numbers of people sleeping on the streets of UK cities in the late 1990s. This had become a visible symbol of poverty and inequality. It had spawned a small industry of shelters and soup kitchens (some of which I had volunteered at since being a teenager). Many thought street homelessness to be an unfortunate but unavoidable fact of life.

To solve it, we created a small and diverse team in central government, analyzed all the dimensions of the problem, developed new ways to measure it (through regular counts of the numbers on the street), explored multiple options, and narrowed them down to a few that looked most promising (which aimed to stem the flow of people coming on to the streets from prison, family breakdown, and the military, and put in place case management to deal not just with their lack of a home but also with the mental health, drug, and alcohol problems that kept them on the streets). A new team set up to act and iterate fast in light of experience oversaw all this.

Out of thin air, we proposed a target for reducing the numbers by two-thirds in three years (which was announced by the prime minister). This target was in fact achieved a year early, and the numbers carried on downward for many years until the policy changed in 2010 (and the numbers started rising again).

Circling and Digging

Making sense of the true nature of the problem was the precondition for both understanding and effective action. In this case, what appeared to be a problem of lack of housing was actually a much more multidimensional one—involving mental illness, addictions, and weak relationships.

Organizations tend to jump too quickly to solutions as opposed to spending time defining the nature of the problem and knowing how granular to make the discussion. So a vital first step for any kind of problem solving is to circle and dig—to circle around the problem and then dig down to its true causes.

Autonomy makes that more likely—giving free rein to different ways of thinking, and creating a kind of commons for the diagnosis and solutions, whether that commons is only within a small group or shared across a whole system. The balance between different elements of intelligence matters—making the fullest use of different kinds of observation, memory, and creativity, and striking a balance between expertise and openness. As the organizational theorist James March put it, "The development of knowledge may depend on maintaining an influx of the naïve and the ignorant."[9] In the case of sleeping on the streets, for example, the organizations most directly involved turned out to have quite a distorted view of the problem, and we needed to bring in others with much fresher perspectives.

For any problem, it's helpful to establish the facts: what the data show, and what patterns of correlation and causation can be pulled out. But data can be misleading; they may be of limited relevance to the problem (there's a temptation to make too much use of data that happen to exist and manage what's measured rather than what matters). And they may be of little use without good hypotheses, models, or priors to discuss. For example, is youth unemployment in a city the result of macroeconomic conditions at the national level, structural conditions, temporary shocks (such as the closure of a major employer), skills gaps, or matching deficiencies in labor markets? The wrong diagnosis will lead to prescriptions that don't work, which is why a fundamental principle of good policy design has been to start with well-defined problems, at the right level of granularity.[10]

Similarly for a government or city, it matters greatly to know if a problem is fundamentally about the overall strategy being flawed, policy being misconceived, implementation being mismanaged or logistical dimensions being badly run, money being too little (or too much), not enough power (to direct or coerce), people lacking skills or motivation, organizations lacking the right skills and culture, a competitor trying to undermine success, contrary competing pulls (such as corruption), and so on. At the very least, disaggregating the problem and analyzing which combination of the list above is involved makes it easier to move toward solutions.

The Hungarian mathematician George Polya suggested one rule for handling apparently intractable problems: "If you can't solve a problem, then there is an easier problem you can solve: find it."[11] In the case of sleeping on the streets, we found at least a dozen smaller problems that all manifested themselves as people sleeping on the streets. Amazon's Mechanical Turk is another tool for breaking complex problems into thousands of smaller parts that can be tackled piecemeal.

Alternatively, the problem may be rethought at a higher level, recognizing a small problem as a symptom of something bigger. As Carl Sagan once wrote, "If you wish to make an apple pie from scratch, you must first invent the universe"—an extreme comment on the connectedness of things.[12]

This stage of circling can also make sense of how difficult the problem is, which will depend on its dimensionality. A problem will be harder to solve if there are many organizations involved, many organizational boundaries crossed, and few shared frames for understanding. Some problems are bound to need a lot of money or time. Some problems are easier to solve thanks to a body of reliable and relevant knowledge. Of course, it matters too if you have the power to solve a problem.

This circling phase can make use of analytic tools—simulations, models, and scenarios as well as analyses of psychology, economics and markets, and political science.[13] But the main aim of this stage is to redescribe the problem in a useful way, and the sign of having completed the circling and digging phase is a clear definition of the problem along with its dimensions and appropriate scale.

WIDENING

If a group or organization can be clear about the precise problems to be addressed, attention can then turn to the tools for addressing it. Without too much organizational energy, smaller and easier problems (using these criteria) can be solved—perhaps by leaving frontline staff and managers or citizens with enough space to get on with it. Or there may be readily available solutions that can be easily borrowed, bought, or copied.

Some of the ideas they'll find become visible for reasons of good communication or fashion more than hard evidence, so we always have to challenge our own tendency to assume that because something is widespread, it must be good. But as a generalization, for any problem there will be a mix of potential solutions that can be easily adapted as well as imperfectly defined solutions in search of imperfectly defined problems.

In the case of sleeping on the streets, for example, we were able to borrow ideas from many neighboring fields, such as pooled budgets, case management, preventive interventions, and city-level partnerships.

If the problem has been adequately interpreted, and there aren't suitable solutions to be adopted or adapted from elsewhere, then the only way to solve the problem is to come up with something new. There are many ways to think creatively, and arrays of devices and prompts that can help any group to generate new options. As Linus Pauling put it, the best way to have good ideas is to have lots of ideas and throw away the bad ones.

The best ideas may emerge in surprising ways. The literature on insight emphasizes how often the subconscious mind does better in spotting connections, patterns, and possibilities. In Graham Wallas's influential 1920s' book on the art of thought, for example, insights come in four stages: preparation, incubation, illumination, and verification, and the illumination stage often happens almost by surprise, after a period of unconscious incubation.

This stage of widening is never really complete. But time and resources cut it short when we have a sufficient menu of options, from a sufficient range of sources.

NARROWING

Where we go next is to narrow and select, applying criteria to possible solutions or sometimes gut feelings. Is there evidence that they would work? Are they cheap? Are they politically viable? Most institutions have many tools for doing the rational versions of selection, from cost-benefit analysis to investment appraisal. This is the comfort zone of rational, analytic bureaucracies.

Narrowing is essentially a probabilistic exercise. Whether explicitly or implicitly, it involves estimating probabilities of success relative to resources required and drawing on knowledge of the past. In theory, it can be helped by more sophisticated theories of causal inference along with structural models that predict how people and organizations may respond to a change in policy.[14]

Narrowing is much harder with novel problems or in the face of genuine uncertainty. Here the default will be to rely on analogy or defer to the incumbents, however badly they have messed up, or to tap into intuition.

The narrowing can go all the way to solutions. Or it can leave a short list of options. Where there is sufficient time and money, it may be sensible to try out more than one possible solution on a small scale, whether through formal pilots and experiments, A/B testing of variants, or more informal tests. In the case of sleeping on the streets, it was quite easy to test different options quickly in different cities and see which worked.

Oddly, although individual institutions are quite good at narrowing, at a societal level there is a surprising shortage of institutions designed to judge what really works and which ideas deserve to be backed with resources. Markets play this role for commercial ideas. But ideas that could be socially valuable often struggle to find support, even when there is strong evidence that they work.

ITERATING

Since any solution to any problem will be imperfect, it needs to be designed with a capacity to learn and improve. Public problems are not like mathematics problems—with solutions that are either correct or incorrect. Instead, they become more effective over time as bugs are fixed and skills cumulate.

There are many ways of ensuring that implementation is better infused by learning—careful measurement, gathering inputs from those involved in implementation or on the receiving end of services, and "quality circles" and other ways of encouraging the people involved at every level to think about improvement.

But our own solutions may also feed the next generation of problems. Like bacteria, problems develop resistance to our cures. So what works in one decade may not in the next, rendering problem solving more of a circular than linear activity.

PROBLEM SOLVING IN TRIGGERED HIERARCHIES

The truly smart city will have explicit capacities and processes for carrying out the kind of problem solving described above—with wide networks to tap into information and ideas, and sufficient depth of expertise to design solutions that will not only work but also fit with the available political, economic, and organizational conditions.

Bringing these points together, we can also see how the city can handle problems through using triage. Relatively simple, predictable, and repeated tasks are dealt with in automated and standardized ways, including the everyday activities of a police officer, tax collector, or teacher. Higher-tier authorities become involved in decisions, or fixing things, when triggered by problems, scandals, complaints, or new information. This leads either to the restoration of good procedures (first-loop learning) or development of new procedures (second-loop learning). Of course, many governments are more likely to produce problems or misery for their citizens and lower-tier authorities. Yet what's described here is the more desirable alternative, not too far from the reality in many places.

GOVERNMENT AS COLLECTIVE INTELLIGENCE

What of national governments? Governments sometimes aspire to be the brain of their societies. They put a head on coins as a symbol of leadership and to remind people that they know more. They like to see and survey, deploy networks of spies listening to conversations of dissent, map and divide up territory, and use an array of statistical devices to measure the population and understand its subtle movements. Governments liked to believe that they could out-*remember* anyone else; the very word control comes from the Latin *contra-rotulare* (against the rolls), referring to the rolls used to keep records, and every state had at its core the archives. Tally sticks were kept to remember tax payments in the British Parliament, and the Domesday Book recorded the tax potential of every home and farm. And of course states aspire to a superior capacity to make judgments, weighing up rights and wrongs.[15]

But the state now has the capability to be something more like a true collective intelligence. In one direction, it is being taken toward the panopticon—able to see, hear, and analyze everything. These are the powers conferred by an Internet that not only carries all communication between citizens but also increasingly carries communication between things—registering where each car is and the temperature of each home. In this world, the challenge for the state is how to keep track of the floods of data pouring into it, and how not to overstep the mark in its eagerness to know and become a monstrous Leviathan. India's Aadhaar project, for instance, has now provided well over a billion people with biometric identities and is generating vast quantities of data to the government.

But there is also a different kind of collective intelligence available to governments: the potential to collaborate with citizens to see, hear, analyze, remember, and create. This takes us to a less familiar history and possible future. That history has seen many attempts to create a more open, collaborative kind of state, such as in the thinking of anarchists and socialists in the nineteenth century, and libertarians in the twentieth. Sometimes these ideas have borne fruit. For example, in 1916, Wilbur C. Phillips proposed the creation of a national social laboratory that would involve people in reshaping public services. The idea built on Dewey's Laboratory School, which had brought participation and experiment to education. Sixteen US cities offered to take part, bidding for the $90,000 on offer, the equivalent of around $25 million in today's money. Cincinnati won and set up a team of eight people in a living laboratory. By 1920, the lab had established a comprehensive public health system and looked on track to pioneer a different model of government. But as so often happens, a combination of factors crushed the idea, from political nervousness to opposition from officials.

A similar project emerged a few years later in the 1930s in south London. The Peckham Experiment tried to promote the health of the public of a fairly poor urban neighborhood by dealing with all the causes of ill health, from housing to happiness and to diet. It innovated a wide range of ways to monitor and learn from experience. But after several years of impressive work, it too was killed off, ironically by the birth of a national health service, whose doctors neither understood nor appreciated the idea of sharing health with the people rather than just curing a passive and largely ignorant group of patients.

We might hope that these would have a better chance of succeeding now. Today's continuing revolutions in digital technologies are changing the options for how government can be organized, with new tools ranging from sensors and machine learning, to predictive algorithms and crowdsourcing platforms. These technologies can amplify the intelligence of every aspect of government—from democratic deliberation to financial planning, disaster management to public health. They make it easier to learn using the dense informational aura that can be found around every activity, including traces, tracks, and comments.[16]

They certainly enable greater awareness, with floods of data pouring in from the public, businesses, and the Internet and sensors. Many are experimenting with ways of tapping into more expertise and public ideas before policy is crystallized.[17] The US federal government's challenge.gov site is a good example that adapted the principles of open innovation to government, and in the five years after its launch in 2010, was used by 70 agencies, visited 3.5 million times, and hosted 400 challenges.

In principle, new inputs can help states be more empathic. This tends to be a blind spot for technologists and enthusiasts for new tools. Yet as Robert McNamara—former head of Ford, the Pentagon, and the World Bank—pointed out, many of the worst mistakes that states make derive from a lack of empathy. Technologies can help to counteract this risk. For example, tools like social network analysis can reveal the reality of relationships, such as who helps whom in a local policing system.[18] Other tools help people to be more creative, or be better at remembering their own past achievements and failures, or orchestrating knowledge.[19]

STRATEGIC INTELLIGENCE

I've had a few chances to implement ideas that aim to make centers of government more intelligent in various countries around the world. Here I focus on one part of the toolbox (my book on public strategy sets out much more): an attempt to address the question of how to ensure that while every part of government thinks, plots, and plans, a shared cross-government view could be developed, bringing to the surface not only the facts, uncomfortable or not, but also the feelings and intuitions.[20]

This exercise was given the bland title of a strategic audit, but points to how collective intelligence could be better integrated into the business of government. In this case, part of the task was analytic—to observe in a fresh way what was really happening, what needed attention, and what progress was being achieved on old problems and what action was needed on new ones. This involved looking at each area of promises and action, and asking whether they were really working. Then we took a different cut, looking at parts of the population, asking how they were faring and what government was doing to them. This revealed some surprises, such as how little was being done for relatively poor middle-aged women. Another exercise looked at possible future trends and opportunities. From this we put together a map of the landscape, or rather a series of overlapping maps—some looking backward and some forward, some quantitative and others qualitative—to spot emerging patterns, such as the rising incidence of isolation among the elderly, extremist Islam, or adolescent mental illness.

But the most important move was to add in emotion and values as well as connect strategy to personal perception. To do this, we interviewed all of the top team separately and anonymously, so each could talk honestly about their hopes and concerns. That included the prime minister and head of the civil service. Some of the interviews went on far longer than expected—therapy for the powerful, who often find it hard to speak openly to their peers for fear that a misplaced comment might be passed on, shorn of its context. The results were then shown to the whole UK Cabinet and senior civil servants to guide priorities. What should be in the party manifesto? What should be the main areas of spending? Where did policy have to change?

I won't pretend the exercise was perfect. Many of the rough edges were polished off. This was just after the Iraq War, and many had a sense of how disastrous that would prove to be. But as an orchestrated method to help the group think as a group, it worked well and was subsequently copied by other governments. It was honest, open, mature, and a lot more intelligent in every sense of the word than deal making behind closed doors. It helped shape a series of strategies that were published and broadly stuck to over many years (including by the coalition government that took over six years later).

It was also a direct implementation of the organizing principles I've described. It created an autonomous commons for the leadership team—one step removed from egos and interests. It deliberately balanced different ways of thinking, from observation to memory. It helped the group to focus, was reflexive and took stock of what had been done, and was oriented toward action.

Unfortunately this was a relatively rare case. Most government is much more like improvisation and much more defined by conflicting interests. But the good examples point forward to a radical possibility where in order to feed collective intelligence, thought processes are visible, conscious, and open to learning. This movement started with freedom of information legislation and went further with the spread of open data. Extrapolating from these, it's possible to imagine how many aspects of government could be opened up to inspection and improvement so as to tap a broader pool of intelligence: using data from many sources, drafting policies and laws more openly, predicting possible effects, and ensuring much more transparent scrutiny of effects achieved. The building blocks for all of these already exist—whether in citizen- and sensor-generated data, open platforms for problem solving, or independent organizations dedicated to scrutinizing what governments do. Nowhere are they brought together into a deliberate design for collective intelligence in government, however.[21]

If openness became a norm, politicians, professions, and agencies would more systematically set out their expectations, showing how they thought this law might work, whether this prisoner would reoffend, and whether this patient would thrive. Making visible what actually happened would then build in the sort of reflexivity that is so vital to intelligence and institutionalize the three learning loops much more effectively.

An element of this is already present in central banks and the use of predictive algorithms in fields like primary care. We know that learning systems rely on visible goals and means as well as taking stock of surprises. It's a spirit opposite to pompous rhetoric and the worlds of spin. But it would evidently be in the public interest. Hopefully it's not a fantasy to imagine that government could, as in the eighteenth and nineteenth centuries, once again be a pioneer of intelligence on a large scale.

- 13 -

Visible and Invisible Hands

Economies and Firms as Collective Intelligence

ONCE UPON A TIME, THE RESTAURANTS in a town could rely on their regulars. The food might be mediocre and the atmosphere dull. But trying somewhere new was risky and could easily ruin a night out with friends.

Now the market is saturated with information. When choosing a restaurant to eat in, you can browse through diner ratings, comparison websites, and public health and safety inspections. Good newcomers find it easier to set up precisely because potential customers can be more confident that the restaurant will turn out to be good. You may be able to learn where the restaurant sources its food, how it treats its staff, and whether it makes a profit. In some towns, restaurants band together to reduce how much food they waste (for example, providing it to homeless shelters) or fund improvements to public spaces.[1]

Food is a good example of how economies are becoming collectively more intelligent, shaped by feedback and smarter decision making. Looking not too far into the future, it's possible to imagine the food economy becoming still more intelligent, offering consumers data on carbon footprints and the pleasure of past customers, allowing restaurants to exclude customers with a record of hogging tables too long, and using artificial intelligence to design menus, measure and track food waste, or plan employee training.

These are instances of a potential sea change both in economies, which would come closer to an ideal of collective intelligence, and economics, which would become much more empirical, drawing on evidence from big and small data in real time, and much less a discipline that deducts theories from assumptions that aren't always well founded in fact. Taking these ideas even further, we might even imagine an economy that could respond to the preferences and wishes of everyone, not just the simple consumption preferences of people with money. Such is one utopian extrapolation of today's modest steps toward collective intelligence.

Visible and Invisible Hands as Metaphor and Description

Ideas about intelligence play an important role in modern economics, mainly thanks to the concept of the invisible hand, which tells how price signals allow millions of consumers, workers, and firms to coordinate their actions. Hayek wrote one of the most influential accounts of the economy as a cognitive system: "It is more than a metaphor to describe the price system as a . . . system of telecommunications. . . . The marvel is that in a case like that of a scarcity of one raw material, without an order being issued, without more than perhaps a handful of people knowing the cause, tens of thousands of people whose identity could not be ascertained by months of investigation, are made to use the material or its products more sparingly; i.e., they move in the right direction."[2] The result is that the price system also promotes "allocative efficiency" as it draws resources to their most highly valued use.

Much more is known now about the limits of this analysis—the asymmetries and distortions of information that make perfect markets so rare.

But in all real economies, including the food economy, invisible hands are at work. What's rather less well theorized is how they operate in tandem with visible hands. The visible hands of large companies, regulators, trade organizations, and research centers influence and guide the invisible hand of markets responding to changing tastes. The visible hands include the companies that orchestrate intelligence across an economy, such as platforms like Amazon or Li & Fung.[3]

Other visible hands include regulators who inspect restaurants, set limits for salts or sugars, or police supply chains to ensure, say, that horsemeat isn't passed off as beef. They become most visible when there's a scandal, like China's many scandals over food contamination, but frequently work less visibly, like the powerful Codex Alimentarius set up by the UN Food and Agriculture Organization to govern what we eat.

The Components of Intelligence in an Economy

Both visible and invisible hands depend on all the elements of intelligence described earlier, and this provides a good starting point for understanding how real economies work. They can, in other words, be mapped in terms of how well they observe, reason, remember, create, judge, act, and learn, and how well they handle first-, second-, and third-loop learning too. Once we do this, the more simplistic accounts that portray economies as brilliant, automatic mechanisms turn out to be of limited use.

Take observation. Modern economies are awash with information—both objective data about outputs, inflation, and trade, and subjective data about confidence and expectations. Even the best-financed businesses, however, struggle to see important facts, such as the true feelings of their customers or staff, the plans of their competitors, or even their actual levels of profitability. The same is true of governments. Indeed, until the 1930s, governments had few ways to observe economic activity. They relied on rail freight volumes or the level of the stock market. Then a series of new measures were devised that came together as GDP, tracking outputs across different sectors. Simply seeing the economy in a new way made possible radically different methods of macroeconomic policy.

As I showed in chapter 4, some of the more recent tools return to direct observation of economic activity—satellites tracking truck movements or the density of electric lights. Other new tools scrape the web to understand the emergence of new industries, since any new firm in fields like software or games will leave many digital traces, while official statistics tend to use categories that lag far behind.

What is being observed also changes. The big shift in recent years has been better observation of intangibles, such as investment in new knowledge, brands, and design—all of which were largely invisible to the twentieth-century indicators. The day-to-day life of the economy is enriched or sometimes flooded by other kinds of data, like who bought what and where. Within the firm, psychometric tools observe employee's dispositions, and sensors track physical movements or the length of breaks.

The interpretation of these data is then divided between machines that make assessments (Is someone seeking a loan creditworthy?), as well as individuals, and teams. Pharmaceuticals, for example, have been transformed by the use of high-throughput screening or the automated analysis of huge data sets of molecules. Finance has been transformed by the use of algorithms to guide trades, sometimes with disastrous effects.

Other layers of analysis and interpretation are then added by investment analysts, journalists, regulators, and policy makers. Seen as a system, what is intriguing is the combination of proliferating intelligent machine tools for purchasing and trading—often with little human input—alongside the spread of expensive and very human functions like strategy consultancy and executive coaching.

Observation has been massively enhanced, but still has many blind spots, such as psychometrics to spot out-of-control managers or disaffected staff, forecasting devices that can predict downturns, and mood shifts in consumer preferences. So too does analysis miss many of its targets. The share of a firm's value that comes from audits has fallen from over three-quarters to perhaps a fifth—the remaining value, which encompasses everything from brand values to the value of staff, is hard to measure in any reliable way.

This is a broader problem of observation that has contributed to volatility—and periodic crashes—in investment. Many of the types of investment that do the most to create value—from patents and information technologies to brands—are treated as regular expenses. Reported earnings bundle together genuine contributions to growth along with one-off gains and losses. The result has been a widening gap between reported earnings and share prices, particularly for firms with a bigger emphasis on technology and science, prompting a search both for more sophisticated measures and a reliance on cruder ones that are harder to manipulate, like cash.[4]

Every aspect of intelligence can be done well or badly. The organization of creativity in business is a good case in point, full of both good and bad examples as well as a surprisingly uneven use of available evidence. The methods used by firms are shaped as much by fashion as anything else—and there are remarkably few instances of firms deliberately testing different methods against each other to see which are most effective.

Memory is in some respects organized well—say, through customer relationship management systems that collate the history of past interactions between a firm and its customers. But several decades of high investment in knowledge management have left few firms confident that they know what they know, and in other respects economies can be poor at remembering, as is apparent when every recession takes investors and managers by surprise.

Firms generally divide the various tasks of intelligence between different units or people, with one team responsible for data, another for creativity and design, and a third for managing records. The firm may use a portfolio of tools for thinking, such as A/B testing for new ways of reaching consumers, experimenting with a few thousand or hundred thousand for one option, with another group testing an alternative approach, to see which prompts the most purchases. Open innovation methods maybe used to tap into ideas well beyond the boundaries of the firm, and behavioral laboratories to test ideas with real customers.

But because the science of collective intelligence remains embryonic, firms have few ways of knowing how well any of these methods work. There are as yet no metrics to tell them how well they're creating, remembering, or deciding, other than the blunt feedback they get from whether their sales or share price are thriving.

Most thus rely on intuition and feel, or subjective measures to determine how well their cognition is functioning. Most of the study of decision making within firms assumes that they act rationally, computing a range of options, and making decisions based on the data available to them and estimates of probabilities of different results that would follow from decisions. Closer study of how real firms behave, though, suggests that they think in different ways, less through the computation of alternatives and more through rules of thumb that are kept until they stop working. Part of the reason is that the cost of computation is too high and it's also expensive to try out new methods. Rather, more automatic responses are built up through experience and passed on to new recruits.

In a famous chapter of his magnum opus, *The General Theory*, John Maynard Keynes wrote about how inherently unknowable the future is, how risk can never be quantified, and how instead people rely on workaday narratives and conventions to help them through. Crises shake these up and generate new ones. Yet no amount of rational analysis can make these either reliable or predictable.

These narratives guide decisions, but they also play another role. To hold a business together it may be important to sustain a useful narrative and help people feel good as well as committed, and too much reflection or skepticism can threaten this. In the economy, as elsewhere, there are trade-offs between collectiveness and intelligence, and it requires hard work by units and organizations to sustain a culture that can simultaneously see the world as it really is while maintaining bonds of mutual commitment.

All these patterns highlight the inadequacy of the old idea of transaction costs that still dominates much of economics. This was meant to explain the boundaries or limits of the firm. According to the theory—which was an advance when first suggested by Ronald Coase in the 1930s—firms choose a balance of external and internal activities depending on the costs of handling transactions. But seen through a broader cognitive lens, it's clear that this view captures only a small fraction of the relevant questions. Every aspect of cognition can be organized internally or externally, and the principles of cognitive economics can help in explaining the costs and benefits of different kinds of thought along with how they might best be organized. For example, where memory is highly tacit and particularly important to the firm, it will be more efficient to keep researchers and product development teams within the firm. Where problem-solving tasks are easily specified and not context specific, it can be efficient to tap into much wider networks of problem solvers, using the methods of open innovation. Specialist data handling, market research, or design firms may be brought in, but always with the risk of losing the ability to understand these well (and know what counts as good) and the linked loss of an ability to integrate.

Recent work measuring creativity in the economy points to how more fine-grained analysis of cognition in economies can be done. In a major study that has subsequently become the basis for redesigning national statistics, every job was assessed in terms of five dimensions of creativity. This provided a rigorous way to measure the numbers of creative jobs in the economy as a whole, which sectors they could be found in (about half were in creative industries like design, and the other half formed a much smaller proportion of total employment in industries like health or engineering), and their links to productivity, profitability, and pay levels.[5]

Other tools can be used to study the digital aura of traces and data that now permeates economies and reveals who buys what and who shares with whom. This aura is costly—in terms of energy and money; it depends on the odd new economics whereby we exchange personal data and attention for free social media; and it's increasingly supported by sensors embedded into the physical things we use each day, like cars or phones.

The explosion of data that has accompanied the Internet makes everything appear more transparent. Yet one of the fascinating features of a digitally enhanced economy is that much of the intelligence of the modern economy remains as obscure as ever, and dependent on private conversations, secret deals, relationships, and favors. A high proportion of the most lucrative investment and of business at the higher end of financial services is driven more by relationships and privileged access to information than by incentives.

Meanwhile, some of the most interesting aspects of collective intelligence in the economy can be found by asking what isn't seen or attended to, such as the demands of the poor rather than the rich, qualities versus quantities, and things that aren't easily monetizable as opposed to those that are.

Economies as Coupled Systems, Loops, and Triggered Hierarchies

All real economies combine both invisible and visible hands. We can usefully think of these as "coupled" systems. Business markets that rely on the embedded intelligence of the invisible hand are coupled with states that act as steerers, regulators, and shapers, as purchasers and providers. States take some responsibility for the overall intelligence of the economy, particularly in crises when the command centers of states and finance ministries become all important. Indeed, modern economies are peppered with institutions charged with steering to rectify the typical errors of the invisible hand. Some exist to protect consumers, some to keep banks solvent, and others to regulate credit or trade. Some grow organically out of the cooperation of businesses, and others in response to abuses. The more developed the economy, the more it is likely to be full of intermediaries of various kinds, some voluntary and some with formal powers. Seen through this lens of collective intelligence, many of the familiar arguments about economic policy—for instance, state versus market, industrial policy versus laissez-faire—look anachronistic. What matters is the combination of each of these, and whether as a result the whole system is able to adapt or not, to think afresh as conditions change.

Most modern economies look much more like the triggered hierarchies described in chapter 3 than the traditional account of markets offered in economics textbooks. A change in prices, a new production technology, or a shift in consumer tastes are all kinds of feedback that can be accommodated with incremental actions made within the scope of existing models. The branch manager or team can mostly handle these.

Then there are shocks or surprises that are more fundamental in nature. The head office may be pulled in to sort out the crisis, either restoring the previous order or instigating new arrangements. Higher-level shocks may force coordinated action by a few big firms or holding companies, or bring investors into the equation. The highest-level responses will almost always involve the state or states collaborating with each other at a global level. In other words, visible and invisible hands depend on and complement each other as well as sometimes compete, rather like our own automatic and conscious thoughts and actions.

Economies as Combinations of Routine, Entrepreneurship, and Innovation

Collective intelligence also provides a fresh way of thinking about how economies generate new options. Every organization tries to find a rough balance between stability, order, and routine, on the one hand, and novelty and adaptation, on the other. Efficiency depends on the former, and survival and adaptation depend on the latter. Most people most of the time apply existing knowledge and concentrate on making modest incremental improvements. But some have to be given leeway to create, imagine, and explore, at one remove from rules and routines. Staff members may be given time to develop their own ideas in internal accelerators or chances to immerse themselves in the lives of their customers. They may be helped by structures—strategy teams, labs, product development teams, or Skunk Works asked to think the unthinkable.

These can become a trap, however, if they think wonderful thoughts but are too cut off from the rest of the organization. And the very creativity that may be necessary to long-term survival can overshoot if everyone is working on promising ideas, but not attending enough to the needs of the present. As I suggested earlier, there can be no optimum mix of first-, second-, and third-loop learning capacity in an organization. It's only in retrospect that we can know enough about the turbulence of the environment to judge what the balance should have been. It should in principle be possible to map a rough relationship between the rate of change in the environment and the need to invest more heavily in tools for second- and third-loop learning. But this is an art not a science.

Similar considerations apply to entrepreneurship. Most entrepreneurship doesn't involve novel knowledge. The vast majority of new businesses are essentially copies—new shops, restaurants, or consultancy services actually look very much like existing ones.

Yet there will always be a smaller group of entrepreneurs who create genuinely new and surprising knowledge. Radical entrepreneurship can be interpreted as second-order learning—inventing new categories for thought and action, or combining other people's ideas into something novel. It's closer to the ideas suggested by Saul Bellow that I quoted in chapter 4, as entrepreneurs spot another reality that "is always sending us hints, which without art, we can't receive. . . . Art has something to do with an arrest of attention in the midst of distraction."[6]

Here again, though, it's impossible to judge objectively what's really effective. Entrepreneurs can be wonderful storytellers, but are often misleading ones. The biggest danger of succeeding—in any field—is the delusion that you understand why you succeeded. But beyond hard work and luck, it's difficult for any radical entrepreneur or innovator to know why they succeeded where others failed. The same radical idea can succeed marvelously in one place and time, and fail utterly in another. Once again, the owl of Minerva flies at dusk, and it's only in retrospect that coherent accounts can explain what worked.

The same messy mix of the incremental and radical can be found in technological change, with continuous disruption as new ideas and technologies emerge (again, frequently from second-order learning) and existing technologies "call forth" other related technologies competing for resources and attention. Within the economy there are temporary equilibriums where intelligence tasks become routinized, but these face constant challenge from new ideas that can be reliably interpreted only in retrospect.

This is why economic evolution is so full of third-loop learning—entirely new ways of thinking that turn out to be necessary to make sense of the emergence of novel economic forms. The history of management accounting, economics itself, business studies, and novel uses of data are all part of this often-hidden story that concerns how an economy knows itself. Some of these tools grow entirely within the market, providing obvious benefits to firms; market research methods that can segment customers, spreadsheets to manipulate business plans, and implants that track the movements of workers are all examples. The use of artificial intelligence to screen biases out of recruitment procedures are another illustration.

But some valuable new tools emerge more slowly because they provide a bigger benefit for the economy as a whole than they do for the individual firm or entrepreneur—for instance, comprehensive maps of skills needs or technology forecasts.

Economies Depend on Commons

These kinds of tools—like real-time maps of economic activity that combine open data, commercial data, and web-scraping—are best understood as commons, since they provide a general benefit. Part of their value is to reduce the risk of error. Big errors in economies generally follow from a failure to manage the key organizing principles of collective intelligence.

The most common ones suppress the autonomy of intelligence. Governments hide uncomfortable facts or analyses (as Greece did in the 2000s, manipulating economic data), and management in big firms can do the same (as in the cases of Enron or Bernie Madoff). Incentives can have a similar effect. If the main decision makers in a system are rewarded for short-term profits, they have strong reasons to suppress uncomfortable facts or analyses, as happened right across the financial industries in the mid-2000s—a problem made worse because these industries had systematically weakened corrective mechanisms, financing political parties in order to secure favorable and more laissez-faire regulatory changes from governments.

Better-shared measures and metrics, using the most advanced technological tools available, can provide a counterweight to these tendencies toward error, and should be financed and governed as commons, insulated from the day-to-day pressures of vested interests.

There's a more general point here. Any system with a living, effective collective intelligence will contain rich informational and knowledge commons as well as much more private and proprietary information. It's hard, if not impossible, to collaborate, trade, and discover without a body of common knowledge and understanding to draw on. Some of that commons is cultural, such as norms of how to deal with strangers or what counts as a contract. Some is informational, like shared data about prices or production patterns, or gossip about who is reliable and creditworthy. Some is embodied in people, like the human capital created by an educational system. And some is formalized in shared standards, such as URLs, bar codes, and ISO standards, providing common languages and ways of seeing.

These commons are fuzzier than pure public goods that are nonexcludable and nonrival—like the provision of defense or clean air. But they are ubiquitous in advanced economies, and often supported by dedicated institutions (such as standards bodies).

The collective intelligence lens may make it easier to understand the role these play by turning the telescope around. What kinds of information and knowledge are relied on? Which of these are paid for? How much, or how little, does knowledge flow? How do whole sectors or firms turn around when they hit the unexpected?[7]

The answers form a subset of the study of economies based more on intangibles. Intangible value tends to spill over beyond the boundaries of the firm; it's harder to control or manage. In economies with a larger share of activity in intangibles, the job of the firm becomes more about harvesting, adopting, and adapting the ideas of others. Firms have to become more collaborative for the same reasons and the bigger ones have to learn skills of assembly as well as production.

Business Committed to Evidence

A more collectively intelligent economy will better track and analyze impacts, including less obvious ones. This is already apparent in relation to carbon, with a mushrooming of new ways to track carbon footprints through supply chains as well as new ways of institutionalizing the reporting of social and other effects.

So far, however, there has been surprisingly little systematic attention to the analysis of other kinds of business impact. With governments, a good deal of progress has been made to better marshal evidence about what does and doesn't work. I've already mentioned the various What Works centers, arm's-length bodies responsible for testing and evaluating the effectiveness of everything from teaching methods to policing. The principle is that openness and scrutiny lead to better results.

But what of business? If it's right for powerful governments to be held to account for their effectiveness, should the same principles apply to big corporations? After all, large businesses are heavily involved in fields that have been central to the debates about evidence in public policy—from prisons and reoffending to health care and education—sometimes selling to individual customers and sometimes to big public purchasers.

We could easily imagine businesses committing themselves to only providing goods and services that live up to their claims, such as creams and ointments that really do reverse aging, foods that really do improve health, and cars that really do reduce pollution. This is already normal in some sectors, notably in the pharmaceutical industry, which has to test drugs before they can be sold. But in most of the economy it's not.

It would be unrealistic to expect every product and service to be evidence based. Yet it would be realistic to expect businesses to commit to finding out if they work, in a transparent way. This sort of approach is likely to matter most in fields where there are risks of harm. It isn't particularly relevant to fashion (although What Works centers in haute couture are an intriguing idea). It isn't applicable to much of retail, electronic devices, or utilities—though all these should be subject to rigorous independent consumer testing, of the kind Which? has done since the 1950s.

Evidence matters most when businesses are providing goods and services that have some of the properties of public goods, are bought with public money, or make claims parallel to those made within public sectors or involve products whose efficacy is a crucial part of their value. Examples include many products relating to health, from brain gyms to vitamins, diet supplements to smoking cessation kits. Other examples include environmental products, from solar panels to hybrid cars.

The food industry might be transformed by a commitment to evidence. In recent decades, a great deal of smart brainpower has been devoted to finding new ways to pump sugar into foods that are then presented as healthy, manipulating brain chemistry to invoke addictive behaviors. Social media have been greatly influenced by equally manipulative theories that aim to encourage compulsive and addictive behaviors, regardless of evidence that they damage well-being.

Financial services are reluctantly coming to terms with the idea that they shouldn't sell products that are harmful to buyers, but no country has yet instituted stronger laws that would make it illegal to sell a product that is known to be against the interests of the consumer. Education is another especially important field, where there is patchy evidence as to what works, and easy opportunities for big firms to take advantage of poorly informed parents or teachers.

The simple idea that business should be more evidence based will not convince everyone. Some will argue that there is no parallel between evidence for government and evidence for business. Consumers can vote with their feet if they don't like what businesses are doing or think their products are flawed. "Caveat emptor"—buyer beware—sums up this attitude. The counterargument is that despite being surrounded by vast amounts of information, the public has little access to reliable evidence. Moreover, the public is often not the direct purchaser of goods and services; so in many contexts, caveat emptor is logically equivalent to contending that voters should take responsibility for judging whether government policies work, not governments.[8]

Ideally, enlightened management and enlightened shareholders would align around the idea that businesses should sell goods and services that actually work for consumers, just as governments should try to implement policies that work for citizens. This would be a critical step toward an economy that was closer to collective intelligence, self-aware, self-critical, and willing to learn.

- 1 4 -

The University as Collective Intelligence

IN SEPTEMBER 1793, THE LEADERS of revolutionary France closed every university and college in the country. They saw them as bastions of medievalism. A year later, the universities were reopened as laboratories of learning based on Enlightenment principles. Three hospital-based "schools of health" were opened in Paris, Montpellier, and Strasbourg in 1794. They aimed to reform medicine along scientific lines, using experiment and observation.

As the chemist Antoine Fourcroy announced to France's new lawmaking body, the Convention, medicine would now be taught differently: "Reading little, seeing a lot and doing a lot will be the foundation of the new teaching. . . . Practicing the art, observing at the bedside, all that was missing will now be the principal part of instruction."[1] This is just one example of what I have called third-loop learning: a change in how a whole field thinks, one of many examples of how specialized institutions for learning have tied themselves more closely to the collective intelligence of a field, connecting theory to practice.

Many kinds of knowledge contribute to a civilization. Some is abstract and wholly detached from practice, and any society, paradoxically, needs knowledge that has no obvious purpose. But seen in the long view, the justification for the university is that it generates and spreads usable knowledge more effectively than other institutions.

The methods that universities use crystallized in the nineteenth and early twentieth centuries. They include the discipline, research project, peer review, and academic journals. These are all, interestingly, self-referential: they validate themselves rather than requiring external validation. But universities have also been sites of innovation to make knowledge more alive or useful, and open to external validation. Bologna, sometimes described as Europe's oldest university (though much younger than universities like

Nalanda in India or Al-Azhar in Cairo), offered its first degrees in law, medicine, and astrology, as a practical alternative to the theological focus of Paris and the monasteries. In the nineteenth century, famous examples including the University of Berlin and University College London saw themselves as more engaged alternatives to the stagnant scholasticism of the older universities.

As new forms of collective intelligence take shape, universities are simultaneously strengthened and threatened. They have been strengthened by growing flows of money, including higher spending globally on research and development, and the rise in student numbers to around 150 million worldwide. But they have also been threatened by both competitors—companies, charities, think tanks, and networks all often doing similar work to universities—and the new tools being used for observation, analysis, and interpretation described earlier in the book.

You might expect this to be a golden age of reinvention as universities grapple with how to redefine their role in an environment where knowledge and intelligence are far more ubiquitous. If you look closely at contemporary universities, however, you find a paradox. Universities are centers for research on many topics. Yet although universities are good at applying the principles of research and development to other fields, there appears to be little, if any, systematic research and development done on the activities of universities themselves. Universities are great centers of intelligence, but less impressive as centers of intelligence about intelligence itself. They are good at organizing learning, but not so good at learning about themselves.[2] Indeed, there are few examples of third-loop learning in modern universities and few people who see it as their job to ensure that this happens.

The very qualities that mark out the best and most prestigious disciplines of the modern era are often missing when it comes to the ways in which universities think about themselves.

Every university worth its salt can point to imaginative programs that are trying out new methods of teaching, student involvement in the community, or interdisciplinary work. Many have been highly creative, such as Western Governors University, Olin College, and Arizona State University in the United States; the University of Exeter, Coventry University, Imperial College, London, and the University of Warwick in the United Kingdom; Aalto University in Finland; the University of Melbourne in Australia;

Nanyang Technological University in Singapore; or Pohang University of Science and Technology in South Korea. But the overall picture is of remarkable conservatism: relatively fixed formats for courses, research, and roles, and a culture that is slow to adopt ideas from elsewhere. While student numbers have grown, the models used by universities have solidified, so that the great majority of new universities across the world adopt similar formats: three-year courses, degrees, PhDs, lecture halls, and the paraphernalia of course notes provided by armies of professors and lecturers.

There have been exceptions. In the twentieth century some went to extremes. The University of Wisconsin's Experimental College was a two-year program with no schedule, no compulsory lessons, and no semester grades, but a common syllabus of liberal arts and real-world study in a collegiate community set apart from the town (which flourished briefly between 1927 and 1932). In the late 1960s, the University of Vincennes (whose faculty included Michel Foucault and Gilles Deleuze) at one point offered degrees randomly to passengers on buses, while Cedric Price's Potteries Thinkbelt proposed a university situated in railroad cars. Yet these stand out as exceptions.

There are some good reasons for this institutional conservatism, including the long timescales of scholarship, and justified suspicion of fads and fashions. But many bad reasons also explain the resistance to innovation. Higher education institutions rarely close down, however poorly they perform. So the creative destruction that creates space for new ideas in fields like business, politics, or even particular disciplines simply doesn't happen in universities. Then there's the inertia of prestige and reputation. Most of the top universities now were top universities a generation ago. They benefit the most from donations and endowments, and they're most likely to attract the best professors or students. That gives incumbents powerful advantages. How well they did a few decades ago may count for as much as how well they do today. Add to these governance models that discourage risk taking and powerful disciplines, many rooted in the nineteenth century, that monopolize power and prestige (helped by strong incentives for academics to publish in well-established journals in well-established disciplines), and it's not surprising that universities are much less cauldrons of creativity than they could be.

So part of the problem is undoubtedly an insufficiency of creative experiment. But the bigger problem is that even when there is experiment and innovation, what's missing is the system: the orchestrated experimentation that we recognize as key to successful research and development, openness to ideas and entrants from outside that again is characteristic of most really innovative fields, and systematic evaluation of which innovations do and don't work. In short, universities do research and development on everything except themselves. They experiment with new ways of creating knowledge or teaching, but don't do so systematically, or with effective synthesis of knowledge, in the way that happens in fields like biotechnology, medicine, or computing.

None of this means that everyone should constantly innovate. On the contrary, there are good reasons why most universities, most of the time, should use tried and tested methods. But it does mean that a minority should be given the means and freedom to experiment.

The recently created massive open online courses (MOOCs) like Coursera, Udacity, FutureLearn, and the Khan Academy are a contrary example. They certainly signal restless experiment and new ideas coming in from outside. On closer inspection, though, some are as much symptoms of the problem as answers to it—yet another reminder that when there is innovation, too much of it is misconceived. Internet technologies are likely to transform how universities work, giving more power to students, encouraging students to learn from each other, and making it possible to target specialist but dispersed groups of students and researchers much more easily. But many MOOCs have ignored decades of experience of what actually works in learning and technology, and are failing in predictable and predicted ways. The world's innovators in distance learning—from Canada to Russia to the Open University—found ways to widen access to learning, and experimented with all sorts of hybrids, tutors and summer schools, peer support, and high production value content. Again and again, they learned that purely online learning requires high levels of motivation and persistence, and that most learners, most of the time, need online materials to be complemented by direct interaction with a tutor or coach along with the encouragement of a circle of peers. Yet the designers of the first generation of MOOCs ignored these kinds of lessons. Nor was much systematic research and development done to improve their designs; instead, they jumped to models that appeared plausible.[3]

The mixed experience of MOOCs so far confirms the absence of a properly functioning innovation system or collective intelligence about the work of universities themselves. It shows that there is little serious orchestration of institutional memory, little systematic research and development into how different variants of MOOC could work, and no well-informed source of funding and investment to support the growth of models that really do deserve to spread.

ANSWERS AND QUESTIONS

Early in the twentieth century, Thorstein Veblen wrote an influential book arguing for a division between the higher schools, research universities concerned with pure knowledge, and lower professional schools concerned with application. Their roles were not only obviously different but also obviously related in a hierarchy that places abstraction and scholarship at the top, and knowledge in the world lower. In this view, only through detachment from the world is it possible to discover deep truths that lie beneath the surface of things. To contribute to collective intelligence, universities needed their own logics, codes, relationships, and communities, and risked being contaminated by too much contact with the outside world.

There were always different views—engineering, agriculture, medicine, and the military all saw knowledge as more integrated with the world. To learn mastery in these fields it was necessary to engage with theory, but essential to develop skills through practical application. That practical application usually meant engaging with live questions, the lived problems of everyday practice.

More recently, this idea has gained new momentum.[4] Where the traditional university is organized around bodies of knowledge—and specialist experts—an alternative idea organizes the university around questions or problems. So undergraduates learn at least part of the time by working on difficult, unsolved questions on the leading edge of science or social innovation. This method has been developed on a small scale in some of the world's most imaginative universities: Tsinghua University in China, Stanford University in the United States, and Imperial Collage London. It has also become a guiding principle in some specialist universities, like McMaster University in medicine and Olin in engineering.

All emphasize teamwork rather than only individual work—an approach to learning that tries to pull in all relevant knowledge, from whatever sources. These methods seem to be good at motivating students, and good at preparing them for the realities of life and work. But they are threatening to the existing disciplinary silos.

This idea of starting with questions rather than answers also points to a possible future for universities as orchestrators of knowledge. The current university model privileges academic knowledge over all others. But it's not hard to imagine a different kind of university that would deliberately track, synthesize, and organize relevant knowledge whatever its source, placing itself more deliberately at the center of assemblies of collective intelligence.

This has for some time been a common theme in innovation. Challenge prizes (like the original Longitude Prize in the eighteenth century and its twenty-first-century counterpart) offered rewards to innovators regardless of where they came from. Businesses have taken a similar stance with the methods of open innovation, which presume that the best ideas are likely to be found beyond their walls.

A parallel approach sees the university much more explicitly as a node in much larger networks of thought and practice. For example, the #phonar (photography and narrative) course at Coventry points to one possible future: a free, online undergraduate curricula that is entirely open to anyone (so that alongside physically present students, thousands more can join in via the Internet), but also encourages experts and professionals from around the world to engage online as critics and commentators of the students' work.

If the task of the university is not solely to disseminate knowledge but rather to orchestrate the gathering of knowledge and insight wherever it may be located, different skills and tools will be needed. For instance, a great deal of work is under way to better categorize and orchestrate the description of skills and capabilities, so that it's easier to find people with the skills needed to solve problems. The university should be well placed to lead in this obvious evolution of collective intelligence. But this would require universities to challenge their assumption that any knowledge that originates from an academic source is inherently superior to any other knowledge.

So what needs to be done? The possible building blocks of a more developed innovation system and more mature collective intelligence system for universities mirror what is found in some other fields. They include funding flows, people, institutions, and processes devoted to the components that are needed in any mature innovation system: discovery and experiment, evaluation, and then diffusion.[5]

That might be a revolutionary step. But it would only be a start. For any serious program for collective intelligence in universities can't avoid the question of purpose. New knowledge is never a good in itself. Innovations can be good and bad. So radical innovation inevitably raises questions about what universities should be for. In their role as teachers are they primarily playing the part of signaling, or are they actually providing their graduates with the knowledge and capabilities they need? How much should they be thought of as servants of the economy, and how much as spaces for critical reflection on how power and money are organized as well as how they could be organized differently? How much should we think of them as citadels, guardians of virtue and knowledge, in a sea of ignorance, and how much should we think of them as embedded in communities? How can they advance social mobility and opportunity—or are they content to serve the elite and educate the children of the students they educated a generation ago?

These are not new questions. A. D. Lindsay, the master of Balliol College at Oxford University, warned more than a century ago that "[the] man who only knows more and more about less and less is becoming a public danger," and that universities needed to supplement technical skills with understanding of politics, society, and values. They had to promote common understanding and not just an ever more precise division of labor.

The best innovation reinforces universities' role as servants of society rather than separate or self-serving institutions, and the greatest periods of reform in the past were always informed by a strong sense of mission—spreading knowledge, opening up society, and widening opportunities.

- 15 -

Democratic Assembly

THE PARLIAMENT OR CONGRESS is the most visible emblem of intelligence as something collective. It is the place where the community talks to itself, and assesses, decides, and acts. It is—at least in theory—a collective brain that tries to synthesize the many millions of individual brains that it represents.

Before the birth of modern democracy, Thomas Hobbes portrayed on the cover of his book *Leviathan* a state made up of many people, and wrote that "a multitude of men, are made one person, when they are by one man, or one person, represented; so that it be done with the consent of every one of that multitude in particular."[1] In the optimistic view, modern democracies represent this multitude not just by incarnating them in a single leader but institutionalizing some of the various elements of intelligence I've described, in creative tension with each other. The legislature is where new ideas can emerge; the judicial branch roots out biases and is designed for skepticism as well as, hopefully, wisdom; the civil service guards memory; and the executive is responsible for integration for action and so on.

Yet democracy is deeply flawed, whether seen as a cognitive system or one that represents the views of the majority. Voters choose parties more because of identities or loyalty than policy positions. They adjust their views to fit their loyalties rather than vice versa. They respond to irrelevant influences such as global recessions or bursts of growth that have nothing to do with the virtues of incumbents, and even in media-saturated societies, their knowledge can be profoundly distorted.[2] Even if electoral systems worked better, their representatives might still be corrupt, ignorant, or just misguided, and use methods for deciding on policies that ignore most of what is known about how to decide well, while public decision making might still be beset by its well-known vices: bureaucratic capture, rent seeking, and logrolling, to mention just a few.

It's appealing to hope that technology might be the answer. But many of the recently proposed reforms to democracy that use digital tools risk amplifying as opposed to remedying these flaws through permanent referenda and petitions. So how could democracy become a good example of collective intelligence, expanding the brainpower of a society not dumbing it down?

In this chapter, I look at how democracy can be brought closer to an ideal of collective intelligence, and show the surprising role played by two comedian-clowns who saw possible futures that were invisible to the political insiders.

For most of the last two thousand years, democracy was assumed to be the enemy of collective intelligence. It was an interesting experiment that had been tried and failed, but tended to lead to mob rule and chaos. As John Adams commented two centuries ago, "There never was a democracy yet that did not commit suicide."[3] Where there were assemblies and parliaments, their membership was therefore kept tightly restricted—to nobles, owners of property, or the highly educated.

But through the nineteenth and twentieth centuries, democracy survived and thrived. It spread less as a single idea—government by, for, or with the people—and more as a cluster of devices and institutions, some of which point in contradictory directions, and all of which are continuing to evolve.

These devices aim to both contain and amplify power. As a historical fact, democracy grew up mainly as a protection—a source of security for people subjected to oppression, taxation, and bullying from the mighty. And so in its modern forms, it developed an array of rules and institutions that are best understood as constraints: divisions of power, transparency, and competition, all to make absolute power less likely. These all institutionalize Montesquieu's comment that "power should serve as a check to power."

But democracy also aimed to amplify and reflect. One study of twenty years of data in the United States concluded that "the preferences of the average American appear to have only a minuscule, near-zero, statistically non-significant impact upon public policy."[4] That's an exaggeration since most democratic nations are governed in ways that are not too far distant from the views of their citizens, and the battery of tools used in democracy to reflect opinion, from polls and surveys to referenda, achieve at least a rough alignment on many issues. Yet voting is not a particularly efficient method for communicating preferences. Indeed, it's hard to explain why so many people bother to vote when their single vote has such a tiny chance of influencing outcomes.[5]

Much of the debate about democracy has focused on how it could better communicate the views of millions of people to the places where decisions are made. There have been many advocates of a utopian, direct democracy, where the people bypass the representatives and the authentic popular brain supplants the imposter in parliaments. Some of the recent devices come closer to this ideal, including online petitions (helped by sites like change.org, or avaaz.org, or ones linked to prime ministers and presidents) or digital referenda.

Their problem is that these present choices as binary, and often do little to educate the public about the nature of choices or why their opponents disagree. In this sense they are linear rather than dialectical and turn subtle, nuanced opinions into crude polarities.

This matters because the great lesson that was learned, or relearned, from the expansion of democracy in the 1990s was that democracy involves much more than competitive elections. It depends just as much on a strong civil society, independent media, intermediary bodies, and cultures of trust and mutual respect.

If we see democracy not only as an expression of popular views but also as a collective thinking process, then different conclusions follow. In this view, the quality of deliberation matters as much as the quantity of people involved.

Theorists of governance have long encouraged this cognitive aspect of democracy, including ever-broader suffrage, ever more open sessions, ever more open debate, and ever more participation of civil society in framing options, but it has come more strongly into view thanks to the ubiquity of digital tools.[6] The term *epistemic democracy* has been coined to capture this sense of democracy as a way of generating knowledge and tapping the wisdom of citizens, not just choosing between alternatives.

One of the main reasons for skepticism about democracy was concern that its forms were bound to be foolish, less intelligent than the best autocracies. Uneducated representatives of the public would lack the education to govern well. Popular assemblies would be swayed by eloquence and emotion. Those with power would use corruption to keep the support of the mob, paying off special interests and allowing demagogues free rein.[7]

Yet the practice of democracy avoided most of these vices, partly because the formal processes of decision making were surrounded by complementary systems for thinking. Democracy allied itself with free media, science, and social science along with a growing civil society. It also made its peace with the professional state, which R. H. Tawney called the "serviceable drudge," which could, most of the time, contain the madder ideas of politicians and parties. As a result, its forms of discussion allowed contention, argument, and the airing of alternatives.

In the field of justice, juries had shown that the people could be trusted, and were often wiser and fairer than judges as well as less liable to corruption and capture. The devices that evolved around juries are a good example of institutionalizing the autonomy of intelligence: the secrecy of deliberation, so that genuine arguments could be had; detachment (dismissing jurors with a stake in the decision or link to the accused); and ensuring no limits on the time needed to make decisions.[8]

They also reinforce the point that the detailed design is all important. More doesn't always mean better. Christopher Achen and Larry Bartels argue that today we too easily succumb to a folk theory of democracy in which the people are always right and the job of politicians is only to follow.[9] Yet there are many instances where more public participation has led to notably worse decisions, and decisions that are more likely to be regretted in the future, from anti-fluoridization votes to plebiscites on taxation. Opening decision making up can easily empower some more than others, and give scope for special interests or the most vocal to out-organize the majority. Democracy works when its detailed designs amplify thoughtful inputs and constrain the less constructive ones, just as juries work well thanks to rules that prevent the crowd from being foolish.

The implication is that democracy needs an ecology of institutions that serve collective intelligence and the quality of what we could call the *democratic commons*. These include institutions committed to the autonomy of intelligence and serving truth. In the United Kingdom, I helped design a cluster of organizations to inject more facts and evidence into the system. They included the Alliance for Useful Evidence with several thousand people involved in generating and using evidence, a dozen What Works centers supported by government, and programs of events at political party conferences, civil service bodies, and charities. Their role was not to force evidence on an unwilling system but rather to counter false claims and ensure that anyone making a decision, from a head teacher or police officer to a policy maker in government, could at least be aware of the available state of knowledge. Similar institutions exist in other countries, including offices of technology assessment and budget responsibility. Their role is to help democracy remember better and so judge better.[10]

Other institutions that serve this commons focus on creativity and possibility, like the Committee of the Future in the Finnish Parliament. Indeed, Finland went a step further, creating the Open Ministry, which allowed the public to propose legislation and comment on ideas. Around the world, many experiments are seeking to make democracy more of a conversation, informed by facts and reasoning as well as emotion, rather than a monologue punctuated with elections.[11] My organization, Nesta, in the United Kingdom helped pioneer one set of tools—D-CENT— which allowed political parties, cities, and parliaments (in Spain, Finland, and Iceland) to involve the public much more systematically in proposing issues, suggesting policies, commenting, and voting.

That we were able to do so owes much to the role played by two comedians whose ideas point to both the potential and limits of new forms of popular decision making. One was Jon Gnarr, mayor of Reykjavík in Iceland. After the disastrous financial crash of 2008 had discredited Iceland's political elite, he set up the Best Party, initially as satire, with a program that included free towels in all swimming pools, a polar bear for the children's area at the Reykjavík zoo, and the elimination of all debt. But he won election to become mayor and turned out to be effective as a municipal leader. Under his watch, the city supported a platform called Your Priorities or Better Reykjavík, which pioneered the new democracy— allowing citizens to promote ideas, comment, vote, and compare hundreds of initiatives. By 2016, it was involving about one in ten citizens in proposing and voting on ideas to spend a multimillion-dollar budget. The site requires that people write arguments for and against options, which means they can't simply abuse people they disagree with. The site then ranks the arguments as well as options, and uses strict and effective controls on trolling and abuse. Similar models have been adopted in many other cities, including Paris, where Mayor Ann Hidalgo allocated a significant share of municipal monies for participatory budgeting of this kind, which in 2016 involved 160,000 citizens voting on 200 options. Their aim is to turn city governance into more of an open, continuous conversation with citizens.

The second pioneering comedian was Beppe Grillo in Italy. Again, as in Iceland, a crisis had undermined confidence in the political elite. Helped by Gianroberto Casaleggio, an Internet thinker, he founded the 5-star movement (MoVimento 5 Stelle or M5S) in 2009 to challenge the incumbent political parties. The party was designed for the Internet era—with members making decisions on what the laws should be, party tactics, and candidates. Over a sustained period, it regularly won 20 to 25 percent of the vote in elections, and won power in Rome and Turin in 2016. Similar models of party organization have been adopted elsewhere, such as in Spain's Podemos party, which won control of several of Spain's largest cities in 2015, and involved over 300,000 people in shaping policy through its platform Plaza Podemos.

None of these models is yet mature. The newer parties are still uneasy about taking on the responsibilities of power. Too many of their tools are more expressive than epistemic, and the more direct forms of democracy are still in competition with the older ones.

Iceland is an interesting case in point. Its national parliament involved the public in an open process to rewrite the constitution, with online inputs and a representative commission. The rewritten constitution was then endorsed in a referendum, yet ultimately rejected by the Parliament after a general election had changed its makeup.

But the tools available are evolving fast, especially in providing new inputs to parliaments and assemblies that may retain the ultimate say. As so many other areas of life have been transformed by technology from shopping to music, holidays to finance—it's unlikely that democracy will long resist, particularly if it wants to retain the engagement of young people brought up in a digital world.

To understand how democracy could become more like a collective intelligence, amplifying the best rather than the worst of a society, it's useful to break the democratic process down into a series of stages, each of which has distinct cultures and requirements, and that at their best, can combine the breadth of open networks and focus of concentrated decision making. These include the following:

- Framing questions and determining what is worthy of scarce attention and through what lens it is to be seen (for example, whether climate change matters and whether it's soluble).

- Identifying and nominating issues that might be amenable to action, like how housing can contribute to cutting carbon emissions.
- Generating options to consider, such as how to retrofit old houses.
- Scrutinizing options (for example, using cost-benefit analysis or analyzing distributional effects).
- Deciding what to do, such as whether to implement subsidies or tax breaks or introduce new regulations.
- Scrutinizing what's been done and judging whether it's working.

That's a simplified account. But it helps to show that direct online democracy should mean different things at different stages. So, for example, the early stages can be open in nature. The online platforms like Loomio that allow for conversation between relatively small groups, or ones like Your Priorities and DemocracyOS for larger groups, play their most important roles in these phases of identification and generation of ideas.

Yet the closer any system comes to decisions, the more accountability matters, the less acceptable it is for participants to be anonymous, and the more important it is to know whether special interests are involved. Some prioritization can be done on a large scale. Difficult decisions involving trade-offs, though, will be easier to make in small groups with clear responsibility. On the other hand, for the final stage of scrutiny, civil society, universities, and independent organizations come into their own.

Democracy can be open and messy, trying many things and then letting experience sift the good from the bad, the just from the unjust. But separating these stages out *can* lead to better decisions. For instance, on many contentious topics it's better to distinguish stages of analysis and option generation from the stage of advocacy. If everyone can broadly agree on the main facts and scenarios before egos are attached to positions, the net result is likely to be a better decision. This approach has been adopted for issues as varied as pension reform and climate change (where the International Panel on Climate Change plays the first role and leaves the job of deciding between alternatives to politicians). More recently, the UK Parliament, for example, has experimented with online "evidence checks" on key issues, tapping into a wider community of experts to agree on the main facts before debating options.

At a national level, Taiwan's vTaiwan process goes a step further. It's one of the more ambitious recent attempts to fuse online deliberation with formal legislation, using a series of stages to air options and issues before turning to formal policies.[12] In a first stage the facts are established, then people are asked to express their feelings, and to proceed, options have to show that they have strong support in terms of both evidence and emotion. Other illustrations of contemporary experiments to open democracy up include the Hacker Lab based in Brazil's Parliament, France's Cap Collectif, which involved twenty-one thousand people in drafting a digital law in 2015, and Portugal's national government experiment with participatory budgeting accessed via ATMs.

Many of these examples reflect the influence of outsiders. The vTaiwan pioneers came out of the Sunflower Student Movement, which had surrounded and then occupied the Taiwanese Parliament in response to a proposed trade deal with China (before its leader, Audrey Tang, became a minister). Iceland's experiments grew out of the financial crisis of 2008 and the ensuing "Kitchenware Revolution." The radical initiatives in Spain grew out of Podemos, itself the successor to the antiausterity 15-M movement. In Estonia, a scandal around party political finance created the conditions for the birth of the Estonian Citizens' Assembly, which in turn led to the Citizens' Initiative Platform, Rahvaalgatus.

All of them aim to connect more people into decision making. The best are both online and off-line.[13] But they all face the challenge of working out which kinds of issues are best suited to what kinds of engagement. Issues involving deeply held beliefs may not be so conducive to thoughtful deliberation. More online public airing can simply raise the temperature. Equally, there are many issues on which crowds simply don't have much information, let alone wisdom, and any political leader who opened up decision making too far would quickly lose the public's confidence.

For example, an issue on which there is widely shared knowledge but strongly contested values (like gay marriage) requires different methods from one that is more technical in nature and dependent on highly specialized knowledge (like monetary policy). A contested issue will bring in highly motivated groups that are unlikely to change their views as a result of participation. New forums for debate can polarize preexisting views as opposed to encouraging deliberation.

With specialized issues, by contrast, wide participation in debate may risk encouraging unwise decisions—which will subsequently be rejected by voters (How much would you want the details of monetary policy or responses to a threatened epidemic to be determined by your fellow citizens?). So some of the most useful tools mobilize more expertise than is immediately accessible to government, and filter out inputs based on opinion rather than knowledge.

The experience of direct democracy has also pointed to many other less obvious lessons. For instance, large-scale conversations need a human face—someone to orchestrate, respond, and synthesize. That may be a mayor, minister, or prime minister, or it may be a figure from the media. Then there's the need to let people know what has been decided and why. Satisfaction with democratic processes seems to depend more on this than on whether our own proposals are adopted. Good digital formats—as with meetings—give space for the quiet and introverted as well as the loud-mouthed, and can increasingly show whose views are representative and whose are not as well as who is well connected to others and who is isolated.

Some issues are deliberately kept at one remove from the fierce glare of everyday politics. Instead, institutions are set up at arm's length from governments and parliaments so that they can achieve a more specialized phronesis. Central banks, regulatory agencies, and science funding agencies are all designed to be accountable on long not short timescales, insulated from the immediate pressures of politics and public opinion.

Meanwhile, for scientific choices that involve both ethics and highly specialized knowledge, open public deliberation may be important to both educate the public and legitimate decisions. Stem cell research, genomics, and privacy and personal data are all issues of this type. The debates surrounding mitochondrial research are a good recent example of successful public engagement, and the regulation of machine intelligence is likely to be an important case in the near future.

It will already be clear that the quality of deliberation matters as much as quantity. The principle of one person, one vote is usually taken to be an absolute requirement for democracy. But many other options have virtues, at least for some of the stages of decision making. Some people have advocated systems that can distinguish strength of feeling; imagine, for instance, if you had ten votes, and could choose whether to concentrate them all on a single candidate and party, or spread them out. Others have attempted to revive the nineteenth-century arguments that recognized expertise should give some people more weight than others. This happens on other parts of the Internet, although it clashes with a basic principle of equity. The general point is that what works best may not be a single system, deduced from absolute principles, but instead an assembly.

This is even more striking in relation to transnational democracy, such as how to help the five hundred million citizens of the European Union feel engaged or how to open up UN decision making.[14] These tend to work much better in the early stages of decision making, like influencing what is on agendas, proposing ideas, and scrutinizing options. They're far harder to make work for the crucial stages of decision making—partly because the numbers taking part are insufficient for legitimacy, and partly because the interaction of public debate, political representation, and mass media that fuels democracy within nation-states is far harder to orchestrate on a global scale.

This overview of recent innovations in democracy points to a fundamental conclusion: the general goal for a democratic system should be to mirror all types of intelligence rather than focusing exclusively on voting. It needs to be able to observe reality well, from economic facts to lived experience, and not be deceived by myths or misled by anecdotes. It needs to be able to reason and consider, using argument and deliberation. It needs to concentrate on the problems that matter, creatively explore possibilities, remember, and then make wise judgments.

Parliaments on their own cannot perform all these roles. Instead, they depend on the strength of their surrounding ecology—media, campaigns, and universities. Politics has distinctive ways of thinking, observing, and remembering. Some of these can work well to generate options. But they can as often dumb down, decrease, and blunt evidence. Competitive politics fuels populist styles that divert attention from difficult questions, encouraging misinformation and the worst kinds of confirmation bias. The need for prospective leaders to prove not just that they have good judgment but also ready-made programs leads to commitments to foolish ideas that are then hard to disentangle. Even the parts of the system that should be most useful can become part of the problem. The Washington think tank circuit, for example, has a turnover well over $1 billion, but a high proportion of its output recycles opinion-based research rather than the opposite.

Political parties should be a better part of the answer. At their best, they are good at understanding public experiences and feelings, orchestrating policy options, and then synthesizing complex issues into comprehensive forms. But most of the machinery of contemporary political parties is devoted to campaigning not thinking, with a few rare exceptions like the German parties (which receive generous state funding to help them think). This is one of many reasons why the party model that so dominated twentieth-century politics is long overdue for reinvention, and why the experiments in Spain, Italy, and elsewhere are at least pointers to how in the future political parties could be more consciously designed to think rather than being designed primarily to act as cheerleaders for a cadre of professional politicians.

For reformers, the role of politicians is particularly challenging. Is it better to have a highly skilled ruling elite, in the way that China prepares its leaders through intensive training over many decades? Or is it better to have rulers who literally reflect the people they serve, including their ignorance? Is it better to promote high turnover, such as through term limits, or encourage a cadre of specialized politicians with deep knowledge of how the system works?

In my view, leaders should be assessed for their readiness for their jobs, should be trained to fill the gaps, and should learn systematically on the job. We should want systems that are sufficiently open that incumbents can be sacked, but not so fluid that they are run by amateurs, which implies a much bigger role for education on the job. In the West, however, this is very much a minority view.

This is one of many questions that may find different answers on different scales. A small city like Reykjavík can run a successful online tool for citizens to propose ideas and comment. There's a directness and authenticity about the points made. At the other end of the spectrum, a nation of three hundred million like the United States or thirteen hundred million like India is bound to struggle with online engagement, since well-funded lobby groups are likely to be much more adept at playing the system. More systematic rules, more governance of governance, and a bigger role for intermediaries and representatives are unavoidable on these larger scales. Democracy isn't fractal; instead it's a phenomenon, like much biology, where larger scale requires different forms, not just a scaled-up version of what works in a town or neighborhood.

Crowds can help with many tasks. But they are especially badly suited to the job of designing new institutions, crafting radical strategies, or combining discrete policies into coherent programs. They are good at providing inputs of data and ideas, but not for judgment.

So the challenge for the designers of new forms of democracy that look more like collective intelligence is to attend to the details and fine grain. What emerges is likely to be a hybrid, fusing representative and direct elements, and honest that the buck still stops with elected representatives. In other words, what will work best is not a pure form of democracy that is deduced from a few absolute principles but instead an assembly that can evolve in light of experience and pass the retrospective test of having amplified the collective intelligence of the society rather than having dumbed it down.

- 1 6 -

How Does a Society Think and Create as a System?

> Not everything that is faced can be changed, but nothing can
> be changed until it is faced.
>
> —*James Baldwin*

HOW DOES A WHOLE SOCIETY THINK from the bottom up as well as the top down? How does it imagine radical new options? And how can a system think as a system?

It's easy to see what constrains radical thought: the pull of convention, the influence of powerful interests, and the inertia that reinforces mental frames that reflect these interests. The more expert someone is, the harder it can be for them to see alternatives—which is why there is such value in concepts like "beginner's mind" or the need to unlearn in order to learn. Knowledge can empower, but it can also constrain, as synaptic patterns become habitual.

Societies do think radically, however. Usually the fertile experiment and innovation happens on the edges, where people dream up new ways of living and working and promote them, or entrepreneurs create new ways of doing business. In rare cases they foment wholesale revolutions. Their ideas are more frequently spread by translators closer to the centers of power who spot, invest, and generalize. Political parties were one way of translating the energy of the periphery into the language of power, priorities, laws, and constitutions. Social movements can perform the same role. During some periods, universities played a big part in generating alternatives, and faced with a difficult challenge, a government leader would appoint an eminent professor to devise answers in groups of experts, advisers, and commissions.

The best of these were consciously critical. Critical thinking looks at what is around us as a construct, plastic, and human-made, without being deceived by its apparent naturalness. It looks at history in terms of what was suppressed or ignored as well as what was victorious and celebrated. It looks at the present in the same way—surfacing other views and neglected interests—and it looks at the future as possibility.

Critical thinking of this kind has transformed how our societies conceive of gender and sexuality, race and colonialism, disability and the environment. The measure of each is that the very process of thinking new thoughts and recognizing new truths can be deeply unsettling.

Each involves rediscovery and deconstruction, and then the hard labor of making and shaping, turning insights into laws or institutions. In one view, that labor can only be performed by the people directly affected, and there were good reasons for radicals to want to amplify the voices of the marginalized, who were all too often noticed, if at all, only as abstractions, categories, or numbers. Direct experience, though, is not a precondition for relevant thought. Throughout history, the labor of social design has been undertaken by many people with no direct experience of the problems they are trying to solve and little authentic legitimacy. But that doesn't stop their work from being useful. The same is true of science and technology.

Each of the elements of critical thinking has a different verification principle. The rediscovery of an alternative history can be verified in the same ways as any other history—there is evidence, writings, artifacts, and conundrums otherwise unexplained. So can the facts of the present: critical thinking can point to its truths being deeper, richer, and stronger in explanatory power than the alternatives. Yet proposals for change cannot be verified in this way. Their only verification comes from practice—from implementing them and learning along the way. No one can prove in advance that a new law or way of running a society will actually work. Instead, the best anyone can do is to assemble elements—experiments, examples, and analogies that congregate together to form a different way of running things.[1]

RADICAL SYSTEMS

I want to apply these arguments to the ways in which systems think as systems. We live surrounded by systems of all kinds that sustain life. These provide us with energy, transport, health care, education, or food. Typically these combine many kinds of organization, and roles, and depend on laws, regulations, cultures, and behaviors.

Think, for example, of the system that supports a frail eighty-year-old in her home, helping her to survive. It will include hospitals and doctors, there to help her when she has a fall or needs a prescription. There may be social workers and health visitors to assist with more everyday needs. Her close family may visit her, buy things for her, and give her emotional support. If she's lucky, she might have specialist services—like "Fixer Sven" in Sweden for simple tasks like putting up a picture or changing a ceiling light, ITN in the United States to provide car rides, or a concierge in her building who looks out for her.

These are a loose system. But they rarely work as a system; they don't talk to each other, share information, or coordinate their help. They are distant from the kind of collective intelligence that might be more useful to her, able to predict her needs as well as meet them swiftly, kindly, and efficiently.

So how could a system think as a system to change itself and improve some fundamental failing? How could it become more of a collective intelligence?

Who Is Served by the System?

The example of an isolated woman in her early eighties living in a fairly prosperous Western city highlights how hard it is for systems to be human. She is likely to suffer from multiple health conditions along with repeated episodes and crises that take her in and out of the hospital. She may lack close friends and family, and there is a good chance that she is "high risk" and high cost from the state's perspective. It's highly likely that she is not happy with her situation, interacting with many formal systems, none of which really understands her. Even when the elements of the system work well—for example, the ambulance comes fast when she calls—the net effects of various optimized elements are visibly suboptimal. Better prevention, better care in her own home, more everyday emotional support, and better quick responses to minor crises would all make her life better. What she needs is a collective intelligence that is good at observation, memory, empathy, and judgment, and that can sustain a knowledge commons about her condition. But these are all difficult for the system to provide.

This problem is a typical one in modern cities. Large-scale administrations have become fairly adept at dealing with standardized needs and tasks, from providing primary education to collecting taxes. They are much less adept, however, at these more complex tasks—ones involving multiple agencies, combinations of physical and psychological needs, and effects that have many causes.[2]

Knowing Itself

A first task is for the system to recognize itself: Who is part of it, and who is beyond it? The simple answer is that the boundaries are defined by whoever can have a significant impact on a definable part of the problem. In this case, acute unplanned hospital admissions could provide a focus (and source of measurable targets, such as, How do we reduce these numbers by, say, 50 percent?). The limits of the system boil down to who recognizes the problem as their own or as something they can contribute to.[3] But there will be fuzziness around the edges, particularly when we recognize the roles of family and community.[4]

Some tools can map who is in the system; social network analysis methods, say, can survey the people involved to ask who is helpful to them or who they receive information from. This creates a more realistic map of the everyday workings of a system, usually at odds with the formal institutions or the view from the top.

Next we turn to identification: What is the problem to be solved here? The state might see her as a problem of delivering care, so that her acute diseases are spotted and cured fast. But what would she say about her own needs? My own experience of interviewing the isolated and frail elderly suggests that they give different accounts than that given by the system, with much more emphasis on support, care, and friendship than clinical treatments.

Civil society and the media play roles here in a constant iteration of claim and argument that turns individual experience into recognized common problems. Social media are increasingly important as social sensors. In the case of the eighty-year-old, the rise of isolation as a recognized problem is a good example, which has moved from being a purely private to a partly public concern, helped by anxieties about the rising costs of elder care and hospital-based systems that pick up the mental as well as physical illnesses associated with chronic loneliness. Often the most important phenomena aren't even captured by any data, especially official data. Loneliness is a good illustration of this; it's not measured and is visible more through the absences than the presences, such as the people who aren't receiving phone calls.

The problem then has to be turned into well-formed questions—susceptible to action. That usually means translating them into recognizable forms—economic, social, behavioral, political, or legal.[5] In this case, we can see the convergence of several representations of the same problem: for the medical profession, it's a problem of ill health; for government and finance ministries, it's a problem of costs; and for the public, it may be a problem of unhappiness.

Next, what is to be done? In some cases, there are available pools of knowledge. There is, for example, a huge amount of clinical evidence relevant to eighty-year-olds with multiple conditions, and a fair amount of evidence on how services are organized, some collected in programs like the Cochrane Collaboration. Yet there is surprisingly little knowledge at the system level that's useful in this particular case; most evidence focuses on specific interventions rather than combinations.

Given the glaring gaps in knowledge, the system then needs to find ways of creating new knowledge to fill these gaps and the formal organization of innovation—generating understanding, finding options, and running trials, either focused on discrete intervention pathways or systems.[6]

External pressures can force the system to act. But mobilizing emotions can also galvanize actions and remind the professionals involved why it matters to them to provide a better service. That's made easier if there's a visible commitment by system leaders (for instance, the leaders of the local government or health service), a sense of urgency (such as targets for results to be achieved in ninety or a hundred days), and peer pressure that encourages a sense of personal responsibility.[7]

Here understanding of cultural dynamics helps. Grid-group theory suggests that all organizations and all systems contain within themselves contradictory cultures: hierarchical, egalitarian, individualist, and fatalist. If any one of these becomes too dominant, pathologies result. So successful systems find ways to mobilize different kinds of commitment—the commitment that comes from identification with and obedience to a hierarchy (for example, within a hospital), feeling part of a group (say, doctors collaborating with patients groups), and incentives. Add fatalism for those who might otherwise resist and you have a recognizable picture of many real systems.[8]

Recent experiments that use payment-by-results methods to reward reductions in loneliness combine all three of the main cultures. Hierarchy backs them up (in this case, the local council and health system), the money raised mobilizes volunteers motivated by the desire to help their community, and the organizations involved are rewarded financially according to measurable improvements on a rigorous scale.[9]

Seen in the long view, most examples of truly systemic change involve mutually reinforcing elements—with technologies, business models, laws, and social movements all pointing in a similar direction. A good illustration from the last generation is the dramatic change in attitudes toward waste, which has left us all sharing responsibility for the handling of household waste and prompted a huge expansion of recycling. That depended on the interaction of top-down commands and laws, bottom-up commitment, and horizontal market incentives. It's possible we may see a comparable transformation of the systems that shape care in old age, using new assistive technologies, business models, everyday norms (about responsibilities to parents or neighbors), and professional practices. China is showing one extreme, using public cameras to identify which children fail to visit their elderly parents, and then messaging them directly (and potentially penalizing their "social credit" scores). Other countries are likely to opt for lighter combinations of peer pressure and incentive.

In any context, however, for the system to work well and truly serve the isolated eighty-something described at the beginning of this chapter, it will need to mirror the patterns found in other kinds of collective intelligence. It will need to create an autonomous intelligence—a commons that maps, describes, and makes sense of what is happening, from everyday experience to more objective facts like the numbers of unplanned hospital admissions.[10] It will need a balanced set of capabilities—to observe, create, remember, and synthesize. It will need to be able to focus—using the individual experience as a focal point so as to force the system to attend to what really matters. It will need a reflexive ability—to learn from individual cases of unnecessary failure and evolve its own thinking system.

None of this is inherently hard. These tasks are far less taxing than the design of an accelerator to discover subatomic particles or the job of sending a spacecraft out into distant galaxies. But far less brainpower has been devoted to solving them. And so we stumble on with systems that are a pale shadow of collective intelligence, and leave millions of lives less healthy and happy as a result.

- 1 7 -

The Rise of Knowledge Commons

It's for Everyone

THE INTERNET IS PERFECTLY DESIGNED for sharing—making data, information, and knowledge free. It is a pure expression of the idea of a commons—something everyone can use and share—and an equally pure expression of an ideal of collective intelligence not controlled by traditional power.

But many of the organizations that dominate the Internet are organized on almost-opposite principles, run as private companies selling access, targeted advertising, and personal data to third parties. These have contributed to enormous gains for consumers along with huge wealth for entrepreneurs and investors. The models adopted by firms like Uber, Facebook, and eBay aren't the only ones available, however, and at their worst, risk degrading the quality of information and knowledge on which any intelligent society depends, with algorithms that manipulate and distort decisions, at worst spreading engaging falsehoods more readily than uncomfortable truths.

Different models are needed, and a networked world needs pluralism, competition, and coexistence of forms. In this chapter, I look at the various commons that can, and increasingly should, support collective intelligence as part of new assemblies to help whole systems think. The main threat to physical commons is overuse. The main risk with virtual commons is underproduction. Here I show how that risk might be overcome.

WHAT ARE COMMONS?

Many of the things we depend on are *commons*—shared resources that are free for anyone to use, like clean air and water, forests and libraries, and much of science. The last few decades have seen the rise of a new family of commons. These are the product of digital technologies that can provide services at zero marginal cost, making them well suited as commons.

The Internet and World Wide Web are examples, as are open-source software and repositories like GitHub. Others exist in fields like health care, pooling evidence, knowledge, and experience. Brain Talk is a good illustration, providing access to knowledge about neurological conditions, and hosting discussion groups and commentary. Many other services provided over the Internet have some of the properties of commons even though they are not organized as such. The cost of one more person using the Google search engine or Facebook is close to zero, and these are offered free of charge and have quickly become shared resources even though they are financed by advertising, turning eyeballs and clicks into money. Other digital services that are more obviously commons are supported by voluntary labor (like Wikipedia) and funded by philanthropy (like the Khan Academy). Some are supported by government or, like the BBC, special taxes.

Previous generations of communications technology also had some of the properties of commons. When they emerged, there was feverish innovation to find new economic models—from tax and licenses to regulated monopoly and various devices to redirect resources to support them (such as Britain's Channel 4, funded originally through a share of the main commercial channel's advertising revenues). Some of the most successful solutions funded the commons as commons—that is to say, collectively and not through individual payments for particular services. What resulted were fairly plural, mixed economies for radio and television. The arrival of the Internet, by contrast, has not prompted such creativity, for complex reasons related to ideology and mental blinkers. Every city center thrives through a combination of the public and private, commercial and civic. The Internet has yet to find its equivalent balance.

So what is a commons? The word is used to refer to things whose value is common or shared, and things that are owned, governed, and financed in a shared way (but not run by the state). In economic theory, the term *public good* is usually used to refer to resources that are nonrival and nonexcludable, like air or defense. One person's use doesn't reduce what's available to others, and by their nature these resources are hard to put boundaries around. The term commons is more often used for resources that are nonexcludable but rival: if I make use of common land for grazing, there is less left over for everyone else, and the same is true of electromagnetic spectrum. On closer inspection, however, the distinction turns out to be blurred—even clean air isn't really nonrival, since my polluting car reduces the quality of air for others—and in the digital world, commons and public goods intermingle. The Internet may appear nonrival, but it depends on costly servers, spectrum, and the like. Similarly, although services like Wikipedia and tools like the Internet are organized in nonexcluding ways, there is nothing to stop digital technologies from being surrounded by paywalls. To this extent, they are less "commons-like" than air or water, but still have many of the properties of commons.

If the first part of the definition concerns the nature of the thing, the second part concerns how it is organized, with commons being owned and governed in common. A traditional commons might be owned by a village, trust, or specific community (for instance, foresters), and traditionally, forests, lakes, or grazing grounds were funded and run as commons—that is to say collectively, or through combinations of collective and individual payment (and a subdiscipline of economics has grown up to understand them, pioneered by Nobel prize winner Elinor Ostrom). Public goods have tended to be funded through taxation and provided by governments, but many have been organized through commons rather than states, such as security services providing protection for an industry or area, or streetlighting. There is no rule that states that public goods have to be funded by governments.

From these definitions we can distinguish at least four related phenomena, each of which I look at in more detail later. The first two clearly meet both of the definitions of a commons: classic natural resource commons, like air, water, and forests, and "true" digital commons that behave and are explicitly organized as commons—like Wikipedia. Two other phenomena overlap with pure commons, but don't fit the second criterion: services that provide some of the value of commons, but are not organized as commons (such as Google or Facebook), which I call *value commons*, and public goods such as broadcasting services financed by taxpayers and states (which may be democracies, dictatorships, or empires that are not in any meaningful sense governed as commons).

THE NEW DIGITAL COMMONS

Informational or knowledge commons are made possible by digital technologies that help us to find information at close to zero marginal cost. The Internet itself is a classic commons. The TCP/IP protocol uses algorithms that distribute resources and prevent overuse of capacity by any one user. Other examples are tools for handling information or knowledge, like search engines. Google can be understood as a succession of commons—copying the links across the whole web and indexing it on shared servers, and then offering a free search engine in exchange for personal data. Arguably the service Google supplies is one of the most successful commons in history, providing a free service universally and greatly helping millions to take part in digital projects of all kinds, even though Google is of course not owned or run as a commons.

Some recent commons-like digital services are platforms for exchange, such as eBay, Alibaba, or Amazon, that are not unlike the marketplace at the center of a town that was a classic commons. Some are sources of knowledge—like open data, academic research publications, or the organization that has taken the name Digital Commons, providing a repository for educational materials. Some are ways of handling identity, like OAuth. And some are technological tools, like blockchains being used to create new types of money and verification without the need for a central authority.

The whole digital communications "stack" can be thought of as a series of layers, each of which has some public good and commons characteristics along with some private ones—from the underlying physical layers offering connectivity, through data, networks, and transport, to applications and services. It rests in turn on other commons: the geostationary orbits used by satellites and spectrum used for cell phones.

ARE COMMONS VULNERABLE?

Commons are often thought to be vulnerable, thanks to Gareth Hardin's influential writings in the late 1960s on the "tragedy of the commons" that argued that users of commons will always overexploit resources because of the absence of property rights. In this view, overgrazing and overfishing are predictable patterns that need to be solved either by a strong state or strong property rights.

But several decades of research showed that these tragedies are not inevitable, mainly because the theory greatly underestimated how intelligently communities can manage shared resources so long as they have plenty of time to build up trust and plenty of chances to talk.

In the digital world there have been parallel anxieties that antisocial behavior, trolling, or criminal activity might ruin the new commons. As with the natural resource commons, though, regulatory and governance rules have developed to constrain many of the worst abuses.

The biggest difference between informational and digital commons and the more familiar natural ones is the nature of the risk they face. The biggest threat to natural commons is, as Hardin pointed out, overuse: depletion of resources because individual users have incentives to take out more than their fair share. But this is not a problem with information that can be replicated infinitely at low cost. Instead, the biggest risk is underproduction—because knowledge and information are hard to enclose, and hard to turn into commodities, they are likely to be systematically underproduced. The many institutions that were created to support patents, copyrights, and intellectual property of all kinds aimed to solve this problem by offering individual rewards in return for sharing. Yet these have all become harder to protect and police in the digital environment where copying is so much easier. That, in essence, is the biggest problem of the contemporary digital economy. Many of the services that would be most useful are simply not viable in traditional competitive markets.

DIGITAL VALUE COMMONS AS BUSINESSES

Most of the recent digital platforms and services that have some of the characteristics of commons have grown up as private businesses, with classic commercial ownership models, including stock market listings as public companies. As a result, they have had to focus their energies on making money and paying dividends to shareholders, and their governance looks unlike traditional commons. Meanwhile, whereas recent generations of digital technology have tended to encourage the maximum flow of information and knowledge, the business models used by these private firms frequently depend on creating artificial barriers.

So when Microsoft sells software, it has to construct ever more elaborate security barriers to prevent copying (though it could have chosen, like Linux, to offer it as an open resource). Netflix or Sky have to turn what could be a commons (like traditional broadcast television) into classic commodities, surrounded by expensive barriers and paywalls. Other commons depend on business models that generate revenue from secondary activities. Google continues to make some 85 percent of its revenues from the sale of targeted advertising (the figure was 98 percent in the late 2000s). Facebook, with almost two billion users each month, likewise depends heavily on advertising, along with revenue share arrangements of various kinds. Some are enormous—like Pinterest, effectively a commons for everything from home decoration to fashion—and some are small, like Ravelry for knitting and crocheting, a for-profit firm funded by advertising, but serving as a de facto commons for a community of interest.

There are also other more traditional business models—such as taking a small cut on transactions (as with Airbnb, M-Pesa, or Amazon). But a high proportion of the business models at the core of the digital economy either depend on reducing the utility of the technology or use an indirect model in which the apparent customer is not in fact the true customer. Google or Facebook rarely articulate their business model to users, presumably because users would feel uncomfortable if they were reminded too often that the company's true customers are advertisers. This is different from buying a loaf of bread or paying to see a film at a cinema. Many media businesses in the past used similar methods; the front page of the *London Times* was all classified advertising until the 1960s, and television viewers have always had to put up with advertising. Yet the extent of the mismatch is much greater now.

To be fair, the new digital businesses had little choice; they opted for pragmatic answers in the absence of alternatives. Google was, notoriously, on the point of being forced by investors to hand over its data to a New York ad broker before it worked out how to do this well for itself, and new digital start-ups are under intense pressure to demonstrate plausible revenues—sometimes from direct payments, but usually from advertising sales or paywalls of various kinds that make their service more exclusive.

The combination of first-mover advantage and network effects has given huge windfalls to the small number of firms that came into new fields with large pools of capital. From there on, the ability to gather more data than anyone else as well as gain economies of scale and scope created barriers to entry for anyone else. Google's chief scientist, Peter Norvig, commented, "We don't have better algorithms than anyone else; we just have more data."

A few digital commons have succeeded with quite-different models. Wikipedia is one of a small number that have been funded by philanthropy and supported by a huge input of voluntary labor. The Creative

Commons is both a legal tool for providing information as a commons and also itself run as a philanthropically funded commons, as is the P2P Foundation and various free/libre and open-source software foundations such as the Apache Foundation. The Khan Academy is another case of a noncommercial approach, offering free access to a huge range of teaching tools and funded by grants, notably from Bill Gates. Campaigning platforms like change.org are different again, run as private businesses dependent on fees from NGOs, but have some of the properties of commons, a modern equivalent of Speakers' Corner, albeit answerable to commercial investors. Transport information services like Transport for London are different too, provided by public bodies as classic commons. There was intensive debate about whether they should charge for their data, but offering it for free prompted an explosion of new transport apps. The BBC has innovated in interesting ways around commons, particularly contributing through GitHub to the development of new software.[1]

But the bulk of economic activity on the Internet follows different principles. Google and Facebook together account for some 85 percent of all online advertising, and have dramatically squeezed other industries like newspapers. To do so, they rely to a remarkable extent on the sale and reuse of personal data, without explicit or conscious authorization. Companies sell a service to advertisers, not users. We are essentially being farmed to provide a benefit to third parties. As one commentator put it, we're in the position of a Wagyu cow that's being massaged to make its beef softer.[2] Like the cow, our interests are very much a secondary consideration.

This imbalance can only get worse as technologies suck up and distribute data on ever-larger scales. Who will own the facial recognition data being generated by every retail store on Fifth Avenue in New York City or the millions of CCTVs? Who owns the data being gathered by smartphones that can identify the people in a restaurant or bar? How relaxed should anyone be not just about being recognized but also about big companies mapping their expressions of guilt or anger?

WHY SO LITTLE INNOVATION IN ECONOMIC MODELS?

The period after the invention of radio offers interesting parallels to today. As with digital networks, there was great uncertainty about how radio could be financed and lots of experimentation. Some people thought that members of the public would finance radio by renting short slots. Marconi experimented with the subscription broadcasting of news. Two hundred colleges in the United States applied for radio licenses to create an educational medium. Policy makers considered "toll broadcasting" and taxes on equipment. It took some time for people to realize this was a one-to-many not a one-to-one medium. Yet before long, the world stumbled onto an array of solutions from advertising spots and sponsorship, to donations and the invention of the license fee, and a wide range of public broadcasters ended up being funded out of taxation with varying degrees of independence.

Other fields have seen great innovation in the funding of commons, including a vast range of tithes and levies to fund the stewardship of rivers, beaches, forests, and parks. Recent examples include business improvement districts, which are levies on businesses in city centers decided by vote to fund improvements to common areas.

The ubiquity of the Internet has led to feverish innovation to discover new funding models, with extraordinary creativity around harvesting of data to finance services, turning clicks and eyeballs into revenue, subscription models, crowdfunding, and micropayments. But little of this effort has gone into the discovery of new ways of funding commons as commons, as opposed to ways of commoditizing privately owned data. And of course, many of the models are highly predatory: aggregation platforms take content from one party without paying them, charge a second party for advertising, and allow a third party to use the content for free. This is roughly what happens when Google digitizes tens of millions of books in a database.

Historians will ultimately assess why these economic models have taken hold. An obvious answer is that entrepreneurs and businesses will naturally do whatever they can get away with. A larger answer, however, has to include both the dominance of neoliberal ideology, which means that any model that cannot pay its way commercially is frowned on, and the influence of Silicon Valley, which tends to privilege quite traditional models of commercial investment (while depending on vast public subsidies for the underlying research and development). We all recognize that although we are happy to pay for gasoline or cans of food, we don't see it as appropriate to pay for the right to vote or charge victims of crime for the help they receive from the police. Public goods are best funded in different ways from commercial services. Yet the makers and shapers of the digital world have largely ignored this fact.

One practical result is that many digital platforms rely on venture capital funding, which tends to push them toward rapid revenue growth and models of behavior that are less collaborative and more predatory (since there is only one metric of success that matters: profit)—a pattern that may often, paradoxically, undermine their long-term sustainability.[3]

The Gaps: What's Missing?

The difficult challenge of turning zero marginal cost products and services into an economically viable model means that many new commons that could contribute to collective intelligence don't exist. They don't exist because they would need to be funded as commons. But no one is willing or able to do so.

Hyperlocal media are one example: news services run at the level of neighborhoods of a few thousand that tell you what's happening. The public in many countries has a strong appetite for reliable information about what is happening in its neighborhood. Yet it's not obvious how such hyperlocal sites can be financed. A traditional answer—and a traditional answer for local news of all kinds—was classified advertising. But that is now the preserve of other commons—Google and Facebook—making it hard for new local organizations to compete. So a commons that is clearly wanted and relatively cheap to supply is systematically underprovided, dependent on ultraenthusiastic volunteers.[4]

Another example is reliable knowledge about health. Good guidance on how to handle an illness or symptoms is clearly of great value. It is also a classic commons, since any evidence-based guidance depends on synthesizing huge amounts of data and knowledge. Parts of this are provided as commons through public health systems like the UK National Health Service and projects like the Cochrane Collaboration. By contrast, the nearly two hundred thousand health apps on the marketplace along with the health information offered in newspapers and magazines are of variable reliability and quality. It's easy to see the value that could be created by a true health knowledge commons that gathered together all reliable knowledge in a usable way, with clear guides as to the strength of the evidence it's based on. But there are few easy ways of funding such a commons—other than directly through governments or philanthropy.[5]

A different kind of gap is the provision of trusted identities for online activities. A truly reliable system for creating and managing identities is a classic commons that creates great value not just directly for individuals but also for the whole society and economy. It's a task that is neither well suited to private companies nor to governments unless they're trusted (though India's Aadhaar universal ID scheme has done well on a large scale).[6]

Useful evidence for fields of practitioners, like teachers, falls into this category too. Despite the vast scale of global spending on education, no one has seen it as their job to provide distilled knowledge about what does and doesn't work to the millions of people working as teachers. The United Kingdom's Education Endowment Foundation is a recent example of an attempt to fill this gap and is doing well, building on the pioneering work of figures like John Hattie. It's a kind of assembly, combining experiment, analysis, and synthesis of knowledge. Other new What Works centers will also try to provide comparable knowledge.[7]

Even more important perhaps is the provision of truthful commentary in the mainstream media. Many organizations have a strong commitment to truth—from the BBC to the *Financial Times* and *New York Times*. Many volunteer bloggers now do investigations of their own. But it's surprising how many media organizations don't place much value on truth and accuracy, and economic pressures often explain this. It's also hard to secure funding for serious investigative journalism. Many recent initiatives have tried to fill the gap. ProPublica is one. The Conversation is another interesting and rare instance of a countertrend: a new kind of commons that is also funded as a commons, drawing on content to comment on events from academics in universities and editing their material using the best methods of modern journalism. It's (modestly) funded through a combination of grants from universities and philanthropy. Yet it highlights the surprisingly fragile economic base of truth telling in the Internet age. Similar challenges face media that aim to spotlight positive examples of social change as opposed to emphasizing disasters.[8]

Even more serious is the design of social media that reward items of information for how far they spread, using algorithms that reinforce people's existing assumptions and social networks. One result is that people are frequently surprised by the stances taken by the societies in which they live—which are at odds with the loud voices they hear in their own echo chambers. Obvious falsehoods—from the pope's endorsement of Trump to conspiracy theories about 9/11—circulate widely without any checking. This, again, is the direct result of business models that treat what should be a commons as instead primarily a means of gaining clicks.[9] The furor over fake news during the 2016 US presidential election may turn out to have been a tipping point—and force Facebook, Google, and others to take this issue much more seriously as well as encouraging initiatives like the Trust Project that aim to build veracity into search engines. The scale of the social media platforms overshadows other media. While Facebook has 2 billion users, the world's largest newspaper—*Yomiuri Shimbun*—claims only 9 million readers, while the combined daily audience of CNN, Fox, and MSNBC in the United States is only 3.1 million. It's not yet clear, however, whether the Internet giants—or anyone else—have yet found a sustainable economic model for truthfulness on a large scale.

Law and its practice is yet another classic commons. While the laws themselves are collectively decided, the interpretation and practice of law are not. Large law firms organize legal knowledge in costly and proprietary knowledge management systems, and sell the products of these systems. For the public, the craft knowledge of individual lawyers or legal advice centers provides some help.

But, law could be organized in dramatically more efficient ways using current technologies, making more legal knowledge available as a commons along with focusing paid-for work on genuinely novel interpretation and advice. All laws and judgments could be made open, in machine-readable form, to allow for the development of software to predict the likely results of cases. Platforms like BAILIE—which provides access to court cases in the United Kingdom and Ireland—are steps in the right direction, but their raw data is closed. Technology could then also be used to open up access to the law; for example, the artificial intelligence–based DoNotPay has won over 160,000 cases in the United Kingdom, and bots are being used to provide cheap legal advice. Similarly, the world of contracts could be opened up, made modular, and potentially founded on blockchain (Honduras, for instance, announced that it would use a blockchain to create a secure land registry, but stumbled on implementation). Again, however, this is a commons that has to be financed as a commons.

One of the most intriguing future commons is the aggregation of transport data in cities. Within the next few decades it will become feasible to transform how mobility is organized. In principle, vehicle traffic in cities could be organized much more like traffic on telecommunications networks. The driver would choose a destination, but the network would choose the optimum route, orchestrating traffic to maximize efficient use of the system. This is already happening to some extent with Sat Nav systems and will be taken a step further with driverless cars. Yet the full benefits of a transformed system would depend on all data sources being aggregated into a commons, presumably with some shared multistakeholder governance and ownership as well as rules on how data are provided and used.

It should already be apparent that commons can be found at multiple levels from the local through the national to the global, and some of the most valuable knowledge commons are global by nature. It should also be clear that these are essential parts of collective intelligence; indeed, any kind of collective intelligence relies on a knowledge commons of some kind, even at the microlevel.[10]

When radio and later television emerged, the new economic solutions mainly came at a national level, with funding through national taxes, license fees, and hypothecated funds alongside advertising and sponsorship. National governments are also likely to play a decisive role in helping fund the new digital commons, perhaps through redirecting some of the flows of advertising revenue that go to the aggregation platforms (this was the model used to finance some public service broadcasting), new taxes (for example, on robots or sensors), or microtaxes on uses of personal identity.

At a global level, there's a strong case for using global commons to fund global commons, such as directing license fees for geostationary orbits or spectrum to support the creation and sustenance of content production or useful knowledge in fields like health.

HANDLING NEW DIGITAL MONOPOLIES THAT GROW OUT OF COMMONS

The twentieth century's favored solution for running natural monopolies was to turn them either into public corporations or privately owned but publicly regulated monopolies—the model used in the United States for utilities like AT&T or in the United Kingdom for ITV. In theory, this made it possible to reap the benefits of monopoly—economies of scale and scope—while preventing them from exploiting consumers. Policy aimed to ensure that these services were provided fairly, at affordable prices, and with reasonable quality, and succeeded well over long periods of time. The understanding of the dynamics of competition in oligopolistic or near-monopoly markets has advanced greatly in recent decades, thanks to the work of figures like Jean Tirole, and has shown how much the details matter.

Other organizational solutions include regulated consumer or employee-owned mutuals. There are charities or trusts (of the kind often used to finance and run things like bridges in the past). There are public monopolies, accountable through democratic representatives (the traditional model for postal and telecommunications operators). And there are hybrids like the BBC or Open University (public corporations with some of the characteristics of trusts). Few of these forms have been used for the recent generation of digital commons.

This hasn't mattered so much during the growth phase of new commercial digital platforms and commons. We all reap the rewards from their work. The faster they grow, the more their costs decline. And so we appear to get a wonderful windfall—free Internet services like Google, and cheap new routes to services like Airbnb or Uber. We get a useful commons at apparently low cost. But in a second phase, economic logic is likely to push all these platforms to ratchet prices up and exploit their monopoly position, creating ever more intense conflicts of interest between the commercial interest and public good. It's possible this won't happen. Yet it requires considerable faith in the altruism of owners, managers, and shareholders to believe this.

PLURALISM AND AVOIDING MONOCULTURE

There is a large place for commerce, venture capital, and advertising in the digital economy. But pluralism requires that these are part of a more complex ecology, just as television ended up with a mix of public and private, and benefited from competition between models, not just competition between private companies.

For now we risk a monoculture—an Internet dominated by only one kind of organization (the listed commercial company), based in only one place (California), and using only a limited range of business models (either advertising or harvesting personal data). That can't be healthy. We risk sleepwalking into lock in: dominant positions that subsequently can't be challenged, and undermine collective intelligence at the very moment when it should be advancing. The lesson of all commons (and that of much of the work of Ostrom, the great analyst of commons) is that they require active dialogue, negotiation, and governance; overly generic rules don't work well, and neither does rigidity.

The twenty-first century could be a great age of new commons. We're in the midst of revolution after revolution in technologies that are founded on the ultimate commons—information and knowledge. But these are being squeezed into organizational models designed for the sale of baked beans and cars. Instead, we need to match the imagination of the technologies with a comparable social and organizational imagination.

PART IV

Collective Intelligence as Expanded Possibility

THIS LAST SECTION PUTS THE ARGUMENT in a broader context of politics and ideas. How will the world cope with radically more powerful machine intelligence? How can we bend its potential and the brainpower of people more to ends that matter, rather than to the violence and trivia that soak up so much machine and human intelligence now?

I then turn to the much larger question of whether it makes sense to talk of an evolution of consciousness and what that might look like. Enhanced capacities to sense, interpret, predict, and remember are already changing how we see the world as well as how we think, and although it sometimes looks as if consciousness is regressing, there are clear patterns of evolution that are taking us toward less parochial ways of being where we situate ourselves in a bigger sense of now and broader sense of here.

Throughout the book I try to steer a course between the glib optimism of the evangelists and the grim pessimism of others who see enhanced machine intelligence as the enemy of all that we hold dear. In this final section, I try to clarify just how much these are now questions of choice as opposed to fate or destiny.

- 1 8 -

Collective Wisdom and Progress in Consciousness

G‍IOVANNI P‍ICO DELLA M‍IRANDOLA'S FAMOUS *Oration on the Dignity of Man* (published in 1486) described humans as beings whose essence is to be able to "fashion ourselves in any form that we prefer," and by doing so, attest to our closeness to God. In the centuries since then, this godlike nature has repeatedly both excited and appalled us, as humans have conquered environments, science, and other people.

At the beginning of the book I described how collective intelligence always seems to bring with it something similar: both a greater awareness of possibilities and greater awareness of risk and precariousness. This is the blessing and curse of knowledge, and takes a parallel form for individuals and groups. It's why we can expect little comfort in a world of radically enhanced capabilities, even though it is a logical evolution for a species defined by its ability to think, whose civilizations have grappled with the existential challenges that come with knowing about death as well as glimpsing transcendence.

Too much of the commentary on the evolution of intelligence is overly confident and linear, and redolent of the misplaced confidence of past generations of elite experts and engineers. Alexander Herzen wrote in the mid-nineteenth century that "history has no libretto," and that is, indeed, its lesson. We cannot assume an onward march toward more connected and thoughtful societies. That's why in this final chapter, I dig deeper into the politics and philosophy of collective intelligence, first of all addressing the questions of politics, and then moving on to the role of wisdom and judgment in collective intelligence.

T‍HE P‍OLITICS OF S‍MART M‍ACHINES

My son once told me that I was too dumb to be a droid. The march of smart machines would soon make me obsolete. He was reflecting the many forecasts now available on the potential impact of automation on jobs. Machines may replace or transform half of all jobs in the next twenty years.[1] Whole sectors could be disrupted, transformed, and turned inside out as much of what robots do—observing through sensors, moving limbs, or providing analytic capacity and memory—continues to be embedded into smartphones, household goods, cars, or clothing, and thus into daily life. In the United States, for example, in many states the most common job is truck driver—a job that is eminently and perhaps imminently replaceable by autonomous vehicles.

But history suggests that change happens in much more dialectical ways than futurology usually recognizes. Changes elicit other changes. Trends generate countertrends. New concentrations of power prompt coalitions and campaigns to weaken them. As a result, simple linear forecasts in which technologies just wipe out jobs are misleading.

This is partly a matter of economics. To the extent that robots or smart tools do replace existing jobs, relative price effects will kick in. Those sectors where productivity dramatically increases will see price reductions, and spending will shift over to other fields that are harder to automate, such as personal coaches, tour guides, teachers, care workers, and craft workers. Their relative price will probably rise (as will that of highly skilled jobs in supervision—making sure the robots work, although these too will diminish over time). Labor markets have proven to be dynamic over the last two centuries, coping with massive destruction of jobs and equally massive creation too. There is no obvious reason why a much more automated society would necessarily have fewer jobs.[2]

We also need to think dialectically about demand. Experience suggests that what we want in a more automated economy won't be the same as it is today. We may well be willing to spend a lot on truly smart robots to serve, drive, or guide us. But automation will also raise the status and desirability of what's not automated. Craft is booming in part because of robots. At the upper end, designer crafts fetch high prices for their imperfections as well as perfections. Handmade is now desirable. So is hand grown. These can now charge a premium where at an earlier stage of economic development they were seen as substandard. We should expect even more of a shift toward valuing people. Face-to-face services are already a lot more expensive than commodities, yet at one time they were cheaper. There is no sign whatsoever, though, that the demand for coaches, trainers, masseurs, and beauticians is saturated. Indeed, to the degree that automation further releases disposable income for other tasks, it will shift the balance of the economy even further toward services and especially high-touch ones.

But the most important reason to think dialectically is politics. The majority of technologists and futurologists implicitly assume that the public is dumb and passive—rolled over by big trends that they have no hope of influencing. Two hundred years of technological revolution should have taught us that technological determinism is always misleading—mainly because people have brains as well as interests. People campaign, lobby, argue, and organize. It may take them some time to get the measure of a new bunch of technologies. Yet before long they become agents rather than victims. That's why the conspiracy view of the spread of robots as the ultimate capitalist dream—an economy with only consumers and no workers—is fanciful.

If no one was paid, no one would buy the products produced by robots. Henry Ford had to pay his workers enough to buy his cars. Similarly, a seriously automated economy has to work out some way of generating demand. In theory, every citizen could become a capitalist and just enjoy a flow of dividends from robot companies that they then spent on consumer goods produced by robots. Or they could rely on handouts from the state. Alternatively, rewards could be concentrated in the hands of a few. The point is that in any scenario, questions of distribution quickly come to the fore and open up obviously political choices.

Any new technology sets in motion political battles over who benefits and who loses. The internal combustion engine prompted new rules and regulations like speed limits and new kinds of provision like public buses. Electricity prompted great utilities and public service guarantees as well as a huge apparatus of safety rules. Robots and ubiquitous machine intelligence will prompt similar discussions not just on regulation and law but also on such issues as whether they should be taxed at comparable levels to human workers or whether anyone should have a right to a robot.

Given the number of variables, it's hard to predict where we'll end up on questions of ownership, privacy, and provision. The most likely result of heightened political argument, however, will be that we demand not just robots to serve us but also to incorporate some of robots strengths into ourselves. Indeed if humans have any sense, they will demand the best of what robots have—prosthetic limbs, synthetic eyes, and expanded memories—so that they can keep the interesting jobs along with the status and pay that goes with them as opposed to allowing these to be parceled out.

Since robots are best thought of as having disaggregated capabilities, we will surely want the best of these for ourselves, with our own brains doing some of the aggregation. That's why a movement for human enhancement is more likely than any kind of "singularity." Such a movement may campaign to enhance us physically and genetically, engineering more optimal alleles to produce higher intelligence in our children and grandchildren, in a possible arms race with artificial intelligence.[3] Some societies may choose to distribute these benefits solely according to ability to pay—and since some of these benefits will be expensive, the cost will separate a digitally and genetically enhanced elite from the rest. But it's more likely that this will become a central question of politics—how to distribute these extraordinary capabilities in ways that are fair and seen to be fair.

The dialectical pattern is something more like the human as thesis, machine as antithesis, and synthesis as augmented human, linked into myriad forms of collective intelligence. So when my son says I'm too dumb to be a droid, my answer is, "Yes, that's true for now." Yet I hope that I'm just smart enough to take the best of what the droid has and that the droid is too dumb to stop me.

How to Create an Assembly

The evolution of collective intelligence could move in a similar direction, with hybrid humans linked up in myriad hybrid assemblies, orchestrating intelligence through assemblies that are sometimes deliberately pulled together, like Google Maps or Copernicus, and sometimes grow more organically, like evidence-based medicine.

At their best, these combine all the key elements of intelligence: they observe things, such as the state of the world's water resources or forests, or the reliability of a town's restaurants; they analyze and interpret; they act as stores of memory; sometimes they include a capacity to create new forms; and always they either organize or feed into systems for making judgments about action, and then learn from the results of those actions.

At every stage they depend on organizing principles, including principles of verification (What counts as a true observation or accurate interpretation?), and serve a community of practice. They can be constructed within an organization—as a microcommons—or on much larger scales where they work best if they are relatively independent from corporate or state power, building outward from questions or live problems.

For now, such assemblies are much rarer than they should be. The most visible ones—like Google Maps or Wikipedia—focus on just one or two aspects of intelligence (observation in the case of Google Maps, or interpretation and memory in the case of Wikipedia). A handful, like MetaSub or the Cancer Registry, combine observation, research, knowledge synthesis, interpretation, and creativity. So do the more loosely networked assemblies of fields like medical science.

These are by nature public goods. But few have a reliable funding base. Google Maps—funded from the surpluses of an extraordinarily profitable firm—is an exception that proves the rule.

The other reason they're rare is the absence of either a discipline of collective intelligence, or a professional well skilled in how to design and run them. But my hope is that the next few years and decades will bring the emergence of a new cadre of specialists in "intelligence design," adept at pulling together the hardware and software, data and human processes, that make thought on a large scale effective. This profession will sit on the boundaries of computer science and psychology, organizational design and politics, business strategy and leadership. It will need a repertoire of skills and tools. It also will need, like all the best professions, a strong sense of vocation and ethos in order to link up the elements of the world's intelligence the better to make choices we would in retrospect be glad of.

The Misallocation of Brains

Its mission should be to make the most of human and machine brains. This takes us to a different political question that will involve collective intelligence about collective intelligence. This is the issue of where societies direct their scarce brainpower. Greater awareness of intelligence and its importance will surely bring this question out of the shadows. It's not hard to map which sectors and activities benefit from the greatest investment in both technologies of intelligence and human brainpower. This shows a huge skew in resource allocation toward a handful of sectors: the military above all (which benefits from over half of all public funding of research and development in the United States, and very high proportions in France, the United Kingdom, China, India, and Russia). Other sectors that are disproportionately able to tap into highly skilled brainpower include finance and banking, a few industries such as pharmaceuticals and computing, and within business, functions such as marketing. These are fields that combine highly competitive environments, high rewards, and high status. Small comparative advantages for one organization over another confer big gains. So it makes sense for any organization to spend heavily on brainpower that may give it a small edge, even though the net effect of every organization doing this is only at best a modest overall gain. Military arms races are the most visible aspect of this—where everyone runs harder to stay in the same place.

No one could plausibly claim, though, that these are the fields most in need of brainpower. Ask any group of the public, or for that matter a group of scientists or politicians, what the priorities should be and you get vastly different answers. Health care usually comes first. Needs for new energy or food, or better education will probably come well above the military, which will be far ahead of finance. For the public, the most crucial question is, How will we benefit? And so a more self-conscious, open, and democratic debate about where brains are directed will surely suggest some redirection. Just as important, it will prompt a redirection of research and development on intelligence itself toward these priorities. The great drive for new intelligent machines today is still primarily propelled by the military and intelligence agencies, on the one hand, and competitive businesses, on the other. The development of new intelligent tools to manage the global environment or improve personal health lags far behind, mirroring the broader distortions in how we fund intelligence.

The Deeper Politics of Collective Intelligence

To a conservative, intelligence is embedded in what survives and what surrounds us—institutions, monuments, habits, and norms, perfected through repetition and adding up to the accumulated wisdom of the ages. Being there is proof of wisdom. Survival is the only test that matters.

The radicals think differently. To them the world exists to be remade through rational thought—with abstractions and blueprints, plans and ideals, from which actions can be deduced, with the collective imagination crystallized in political parties, movements, and intellectual currents. What is, the world around us, cannot be the best possible world.

Engagement with the ideas of collective intelligence indicates a synthesis that challenges both of these stances. It suggests that societies evolve best through imaginative trial and error. *What is* is most certainly not the best, and is highly unlikely to be the best for those who fate has born poor or powerless. But what *could be* will never be born fully formed; it needs to be tested out, honed, and recast through experience. In other words, progress has to be incubated in a dialectical way, through praxis not pure intellect, and part of that involves continually questioning the abstractions of theory—abstractions like "the market," "the state," or "society."

Together these points suggest a political stance that leaps beyond the conventional poles of enthusiasm, on the one hand, and fear, on the other. Instead, a more mature politics would fight for wider access to the tools of intelligence and a better allocation of those resources to the things that matter. It would demand that our institutions unleash the full potential of intelligence of every individual and group, mobilize their productivity in the economy, tap their brains in democracy, and expand their agency in private life. That project would be an alternative to the trends of growing inequality along with a widening gulf between a mobile, technologically empowered elite and more passive, impoverished subordinate class. Its project would be the greatest agency of the greatest number.

The Evolution of Wisdom

The deeper political question raised by this book is whether there is a potential for genuine progress in intelligence, and not just a rise in processing speeds or machine learning capabilities. As I've shown, the test of this would be whether humans, separately and together, are able to generate and make better choices. This is as much a question about wisdom as it is about science.

The Canadian poet Dennis Lee once wrote that the consolations of existence might be improved if we thought, worked, and lived as though we were inhabiting "the early days of a better civilization."[4] We don't find it too hard to imagine continued progress in science and technology—ever-greater knowledge of the micro and macro, from the body through to outer space, as well as new materials and faster transport. We can extrapolate from the experiences of the last century toward a more advanced civilization that simply knows more, can control more, and is less vulnerable to threats.

But comparable advances in other kinds of knowledge are harder to picture. The study of wisdom in different civilizations and eras has confirmed that there have been surprisingly convergent views of what counts as wisdom, the highest level of intelligence. These include the ability to take a long view, integration of ethics into thought and decisions, and attention to the details of context rather than simple application of rules or heuristics. Wisdom has generally been taken to mean an avoidance of fixed methods, rigidity of mind and practice, and ways of thinking that clash with the nature of the moment.

A civilization that was wiser would combine the universal knowledge of science with much more context-based knowledge and might even have managed to devise machines that could help people to be wise in this sense—for example, showing the possible long-term effects of decisions or making ethics visible.

Perhaps the most fundamental lesson from the study of wisdom in the past—a lesson with obvious implications for wisdom in the future—is the idea that wisdom entails transcending the boundaries of self or identities and belonging. It involves standing in the shoes of the universe. The very collectives that are containers of collective intelligence thus become transitional devices, useful for a time, but only to be discarded, or at least held on to more lightly. We could imagine that any advanced collective intelligence would combine awareness of itself as apart, a self existing in time and space, and having its own interests, with a sense of itself as part of a larger whole, to which it may owe obligations. This is how wisdom is often seen today: we recognize it in leaders who recognize that their firm is part of a sector and economy, their city is part of a nation and the world, their military organization is part of a larger system that may preserve peace, and their individual and community depends on the biosphere.

The boundaries within which collectives operate are in part necessary illusions that provide a container for thought, a framer of options. But they are understood best as means rather than ends, and what counts as high intelligence is an awareness of how not to be trapped by them. The notion of *holonic* mentalities or *monads* is relevant here. These are imperfect words invented to describe parts that also reflect the properties of the whole, and have simultaneously individual and collective qualities. Most of us are probably more like this than we are like the classically self-sufficient individuals of political and economic theory.

Yet there is an inescapable tension between the pull of what's close at hand and present, and the aspiration to think and act bigger, wider, and over longer timescales. Indeed, many may feel uncomfortable with the idea of a more integrated and open collective intelligence that dissolves the boundaries of the self—that will know us all too well, spot our weaknesses, and challenge our illusions of permanence. Social media are already nurturing a generation more at ease with much of their life being open to others' gaze, with all the advantages and anxieties that brings. A truly well-informed society might be harder to resist than an oppressive state. And certain kinds of collective intelligence could threaten creativity and iconoclasm in the way that too much participation in social media already seems to undermine the ability to be original.

A more optimistic view would expect us to learn the cultural habits of being part of a collective intelligence—better able to share, listen, or take turns. It would hope too that we can collectively learn the wisdom to cope with opposites—to understand suspicion as necessary for truth, fear for hope, and surveillance for freedom.

It's tempting to link possible future evolutions of collective intelligence to what we already know of evolution in the past. John Maynard Smith and Eörs Szathmary offered one of the best summaries of these processes when they described the eight main transitions in the evolution of complexity in life.[5] These were the shift from chromosomes to multicellular organisms, prokaryotic to eukaryotic cells, plants to animals, and simple to sexual reproduction. Every transition involved a new form of cooperation and interdependence (so that things that before the transition could replicate independently, afterward could only replicate as "part of a larger whole"), and new kinds of communication, ways of both storing and transmitting information.

It's entirely plausible that future evolutions of intelligence will have comparable properties—with new forms of cooperation and interdependence along with new ways of handling communication that bring with them deeper understanding of both the outer as well as inner world. The idea of an evolution of consciousness is both obvious and daunting. It is obvious that consciousness does evolve and can in the future. But social science fears speculation, and much that has been written on this theme is either abstract or empty. We see in films and novels visions of machines with dramatically enhanced capacities to calculate, observe, and respond. They may be benign or malign (they're more interesting when they are evil), but we can grasp their implications when we see them scanning emotions on faces, shooting down swarms of attacking missiles, or manipulating complex networks to direct people.

Yet it doesn't take much reflection to remember that throughout history, changes in the quantity of calculation or intelligence are always accompanied by changes in quality—changes in how we think as well as what thinking does. These brought us new ways of seeing the world, such as the idea of a world ruled by scientific laws, not magic, people as sovereign citizens, humans depending on a global ecosystem, or selves as composite, contingent, and partly illusory.[6]

It follows that any future changes to intelligence will similarly combine quantity and quality. History tells us of just such transitions: the passage to larger scale and more comprehensive forms of state; the rise of civility, with cultures able to interact with strangers in cities; and the reductions in violence, as measured by mortality statistics and also everyday encounters.

Even if it is not linear, inevitable, or predictable, there is evidence of a general trend to more communication and what may best be described as mutual intelligence on a larger scale, which tends also to mean more shared rules and protocols, more empathy, a willingness to see boundaries as conditional, and a move away from magic or fate as explanatory. These have been helped by wider access to literacy, communications, and other means of thinking, and by the spread of social institutions committed to open debate that promotes the ability to comprehend whole systems as well as see the interconnections between things in space and time.

There have been many attempts to place these into neat sequences of periods. This was a fashion in the nineteenth century (from John Stuart Mill to Karl Marx), and was advocated by figures like Walter Rostow in relation to stages of development as well as more narrowly in relation to consciousness in the work of figures like Ken Wilber and Clare W. Graves in the late twentieth century.[7] Usually authors place themselves at the most advanced end of evolution (oddly they seem to lack the humility to suggest an evolution well beyond their own brilliance).

These are appealing in their simplicity and often achieve a rough fit with the ways civilizations talk. But any more detailed reading of the historical evidence casts doubt on the neatness of these periodizations. They overlap, and the direction of change is not linear. The absolutist monarchs of the seventeenth and eighteenth centuries in some respects marked a step backward from the sixteenth century. The totalitarian tyrants of the twentieth century marked a step backward from their predecessors, using the technologies of the time, though in order to concentrate power and extend it on a larger scale, but in an entirely vertical, hierarchical way. The causal mechanisms are wholly unclear, since we are biologically identical to our ancestors of a hundred thousand years ago. What's known about epigenetics may explain why different contexts produce different kinds of people and cultures. We just don't know, however.

Just as troubling for the theorists is the problem that so many of the writings that best exemplify a higher level of consciousness date back more than two thousand years. More recent thinkers have not surpassed the wisdom and insights of the Buddha, Jesus, and Lao Tzu. Indeed, almost every other area of human intelligence—science, art, and literature—has seen cumulative progress, but not this one.

A similar uncertainty applies to individual humans. Again, there have been many attempts to spell out a single developmental route for people, composed of predictable sequential stages. Famous theorists of linear development include Jean Piaget, Abraham Maslow, Lawrence Kohlberg, and Jane Loevinger, who all wrote about the staged development of cognition and values.

They point to important truths. But they don't neatly fit with each other, and the evidence is fuzzy. All these theories say as much about people's desperate desire to find patterns as they do about the patterns themselves.

So could we imagine and even fantasize about a more advanced collective intelligence, perhaps one that had transcended the illusion of self and its artificial boundaries, seeing thought as something that comes through us more than being invented by us, a world where the auras of active intelligence are visible in places and conversations, providing a feedback and commentary on the world that's no longer just in our heads but also in between us? Could we imagine a world where our minds and senses are meshed into machine intelligence, with every aspect of consciousness potentially amplified, guided, and linked up? How could we explore the landscapes of inner space, as human brains and digital intelligence combine to generate new kinds of consciousness?

No reliable theory for the evolution of consciousness is feasible, since it would emerge from a culture and consciousness less developed than the one it attempted to explain, and could only be proven right or wrong over long periods of time. But it is possible to imagine, explore, and promote forms of consciousness that enhance awareness as well as dissolve the artificial illusions of self and separate identity.

Such a prospect would terrify many. Yet so would any more advanced forms of consciousness. Better to think, with William Butler Yeats, that "the world is full of magic things, patiently waiting for our senses to grow sharper."[8]

AFTERWORD

The Past and Future of Collective Intelligence as a Discipline

ST. AUGUSTINE URGED THAT "DEVOUTLY confessing that you do not know is better than prematurely claiming that you do."[1] Collective intelligence is just such a field, as fascinating for what's not yet known as for what is. But there is a vast literature to draw on that provides pointers. That literature ranges from the limitlessly broad to highly specific. It fires off in all directions without ever quite coming down to land. It's hard to summarize, not least because there are few shared concepts or frameworks in use. There have been some recent attempts to create a more unified field. For the most part, however, each discipline comes to the topic fresh, or to put it less kindly, from within its own silo. Each provides useful resources for an emerging field of collective intelligence that will offer a more rigorous framework for understanding thought on a large scale. Yet we are some way off from a durable hybrid that all of them can in turn draw on. Here I summarize a few of the strands, which range from the exuberantly abstract to the intensely practical.

A BRIEF OVERVIEW OF THE COLLECTIVE INTELLIGENCE LITERATURE

At the broad end, the Russian mineralogist Vladimir Vernadsky stands out. He suggested that the world would develop in three stages. First came the geosphere of inanimate rocks and minerals, then the biosphere of living things, and ultimately a new realm of collective thought and consciousness emerged that he (and later the French theologian Teilhard de

Chardin) called the "Noosphere." In a similar vein, H. G. Wells wrote
of a "world brain" emerging out of networks. The phrase collective intelligence appears to have first been used in the nineteenth century by a doctor, Robert Graves, referring to the advancing state of medical knowledge,
and separately by a political philosopher, John Pumroy, to refer to popular
sovereignty.

More recently, many others have toyed with metaphors of a collective
brain or mind. Marshall McLuhan in his book *Understanding Media: The
Extensions of Man* provided one of the frames for doing this by describ-
ing technologies as extending our senses and also linking them up. Peter
Russell took this argument a step further in his book *The Global Brain*,
comparing the interactions of neurons in a brain and the interaction of
people and organizations connected through mass media and networks.
Gregory Stock's *Metaman* describes human culture and technologies as a
planetary superorganism, capable of dealing intelligently with problems
that are common to all humanity. Pierre Levy wrote an influential book on
collective intelligence, seen as an aspect of cyberspace, in the early 2000s.[2]

An equally ambitious attempt was Howard Bloom's work on group
brains, which drew on portrayals of collective intelligence in bacterial colo-
nies and insects, and attempted to show the parallels with human societies,
applying the complex adaptive systems and genetic algorithms developed
by John Holland. The aim was a meta-account of how intelligence emerges
and links up.[3]

These ideas are undoubtedly suggestive. But all these writers struggled
with definition and boundaries, and in all of them it is never quite clear
what is being either claimed or refuted.

At the more concrete end of the spectrum, computer science has so far
made the most use of the term collective intelligence, using it to refer to
the ways in which groups collaborate in developing software (for instance,
Linux), orchestrating knowledge (for example, through Wikipedia), or
creating new ideas.[4] Douglas Englebart, one of the pioneers of human-
computer interactions, talked about collective IQ. Others built on this
to address the cooperation of ants, or groups of computers and robots.
Eric Raymond's book *The Cathedral and the Bazaar* provided a central
text, celebrating the ways in which open-source software harnessed many
minds without hierarchy or property rights. In a similar vein, web science,
coming out of computer science, presents the Internet as a vast experiment

in collective intelligence, requiring new concepts and empirical research, such as to understand what motivates people to contribute to the gift economy of projects like Wikipedia. The influential book *The Wisdom of Crowds* by James Surowiecki supplied a common language and defining idea: that large groups will think more accurately than even apparently expert individuals.

The many currents of network theory include the work of Stuart Kauffman and the economist Thomas Schelling, through to the work of Harrison White and Mark Granovetter on the sociology of networks. Where the natural scientists emphasize common patterns in multiple fields, seeing links in quantitative terms as the communication of bits, the sociologists see them as relationships full of meaning and trust. In some areas they overlap, such as in the theory of preferential attachment—that the probability of a node finding a new link correlates with how many links it already has—or the study of homophily—how people link up with others who are like them.

On the boundaries of computer science and euphoric speculation much has been made of the possibility that at some point, digital media will not only link up with each other but will also become some kind of superintelligence, capable of superseding humans. This notion of a coming singularity was coined by Vernon Vinge and enthusiastically popularized by Ray Kurzweil. Like the promoters of artificial intelligence (Marvin Minsky encouraged us to think of networked intelligence as a "society of mind"), these writers tend to predict dramatic changes around twenty years from the present and have been relatively unembarrassed that few of their earlier predictions materialized.

These traditions have greatly influenced contemporary culture as well as helped people to think about large-scale processes for sensing, analyzing, and spotting patterns. Their weakness has been a tendency to rhetorical excess along with the lack of much interest in how people and real-life organizations actually work. As a result, a set of ideas that have their origins in the rigors of computer science have mainly been influential as loose metaphors.

At the other end of the spectrum, a large industry has grown up concerned with much more practical aspects of collective intelligence. These are the sellers and providers of management information systems, data management, and mapping and mining; decision-support tools; and consultancies concerned with creativity, innovation, and change. These are sometimes better at promise than reality, but try to meet the pressing needs of many organizations that know that handling large quantities of data and information intelligently makes the difference between survival and failure. The best are figures like Ikujiro Nonaka, with his theories of the knowledge-creating organization, the traditions of W. Edwards Deming and Chwen Sheu concerned with variation, improvement, and systems thinking, and tapping employee insights, or Peter Senge's concern with holism. The decision theories of figures like James March have also been influential.

Somewhere in between the abstract and practical, a good number of other disciplines have attempted to make sense of aspects of collective intelligence, though they rarely talk to each other, and most don't use the term. They include biologists interested in how intelligent behavior can be found among simple organisms responding to environments, cooperating and flocking, and showing signs of "emergent" intelligence. Much has been learned about how cells come together to form organisms, how organisms (like bees or ants) come together to create complex societies, and even how ecosystems regulate their own metabolisms. The patterns whereby organisms learn how to navigate an environment—for example, bacteria learning how to travel toward sugar sources—can be seen as having close parallels with how cultures learn.[5]

Political scientists have always been concerned with how big institutions make decisions, and have their own repertoire of ideas for making sense of the sane and mad. One of the most interesting recent shifts has been toward a reinterpretation of democracy as a way to tap the good sense of large publics, making judgments on complex choices (the new field of epistemic democracy).[6] Collective wisdom becomes one of the tasks for a political system—how to make the most of brainpower of all kinds to solve systemic problems or mundane ones. Another tradition has studied the dynamics of collective action—and why it may be so hard.[7]

Economics has long had an interest in information, from the perfect equilibriums of Leon Walras, to Ronald Coase's theory of transaction costs that attempted to show why some types of thinking would happen within firms and others outside them. Much of economics has studied information patterns, asymmetries, and decisions within markets, focusing on them as ways to aggregate the intelligence of consumers and entrepreneurs, using the metaphor of the invisible hand (from Adam Smith to Friedrich Hayek and beyond, to Benoit Mandelbrot and Joseph Stiglitz). A more recent current has been concerned with the economics of knowledge as well as the knowledge embodied in firms or sectors, picking up where Karl Marx's theories of general intellect left off in the nineteenth century.

There are traditions in history concerned with how different societies have created, gathered, and spread knowledge (for example, Ernst Mokyr's work on the Industrial Revolution, and the tinkerers who linked mathematics, engineering, and commerce). Powerful traditions in sociology pioneered by Gabriel Tarde emphasized group mind and collectives well beyond the aggregation of individuals, looking at the twin impulses of imitation and innovation. Tarde's rival Émile Durkheim has fared less well in recent years, perhaps because he situated the collective at the level of the whole society, but his theories too were attempts to understand forms of macrocognition: how a whole community thinks, and how apparently individual choices are better understood as manifestations of collective beliefs. His near-contemporary Robert Michels is most famous for his classic work on oligarchy. But his work is really about *Gruppenlebens* (group life) and the trend for all groups to create institutions that then displace collective goals with self-preservation. More recent work that offers ideas relevant to a future discipline of collective intelligence includes Felton Earls' on collective efficacy—showing how a sense of mutual obligation and weness helps to keep crime levels down.

Anthropologists also made progress in addressing "how institutions think," even though the bolder attempts, like Claude Lévi-Strauss's program to create a science of how societies thought, were less successful at translating intellectual fireworks into lasting insights. Mary Douglas's work looked at the "cognitive process at the foundation of the social order" and how "the individual's most elementary cognitive process depends on social institutions," and has proven particularly useful.[8] What resulted were powerful tools for understanding the everyday life of organizations that combine hierarchy, competition, and community in often-uneasy balance.

Psychology has offered its own assessments of how groups and meetings think. The collective unconscious of Carl Jung sat alongside ideas about crowd minds, including Tarde's theories of the interpenetration of individuals and groups. More recently, much work has been done on the boundaries of psychology and computer science, such as by Thomas Malone.

Sandy Pentland has revived earlier attempts to create a social physics that will achieve a predictive, computational theory of human behavior, as has Dirk Helbing with his work on large-scale systems.[9]

In the realm of theory, Bruno Latour's "actor network theory"—not quite a theory, but an approach that encourages seeing humans and machines as inseparable, all playing roles in networks—fits well with the contemporary digital era. A bridge between the computer theorists and other fields is Christopher Alexander's work on the patterns of daily life, which has influenced the design of buildings and computer software (for instance, Wikipedia).

Finally, many offshoots of philosophy and psychology have something to offer, from the engagements with cyberbeings of Andy Clark to Nick Bostrom's work on superintelligence, which warns of the risks of creating machine intelligences that will have the means and perhaps motivation to subjugate humanity. In a different vein, the burgeoning field of wisdom studies has become interested in the past and present analysis of wise judgments along with their patterns, showing remarkable consistency across time and space.

There are many related disciplines that analyze how trust, identity, group bonding, polarization, and so on, affect the ability of groups to function and think. Emmanuel Todd's work on the links between family structure and ideology remains one of the most impressive attempts to show how a social structure shapes how a society thinks (say, the patriarchal family fostering an authoritarian worldview or egalitarian inheritance practices encouraging egalitarian ideologies).

The studies of collective intelligence that come from these disciplines are insightful, and I have drawn on many of them.[10] But the disciplines summarized above all remain disparate fields without shared concepts, causal mechanisms, or in most cases testable hypotheses.

Indeed, they include many contradictory claims. Economics sees intelligence in the invisible hand, largely existing between organizations, and organizational theory locates it within the firm. Theories based on methodological individualism treat the individual as the only meaningful unit, while others see collectives as having their own character, personality, interests, and will.

It's possible that sustaining separate disciplines with distinct and incompatible concepts is the most effective way to deal with a subject as vast as collective intelligence. Undoubtedly many will be keen to police the boundaries of their disciplines. Yet significant advances may be discovered if we look at the patterns that cut across these different disciplines—new insights that can enhance understanding and action.

An Emergent Discipline

To thrive as a distinct discipline, collective intelligence will require a vision that includes advances in machine intelligence but also goes far beyond technology. It will require better theory—ways of thinking about thinking that can integrate what we know about collectives (such as what enables groups to cohere, think, decide, or act in harmony) and what we know about intelligence (covering its various dimensions from creativity to calculation to judgment, and its various hierarchies and registers).

Some elements can be adapted or adopted from existing disciplines. Economics could evolve new concepts of cognitive economics, addressing the costs and benefits of different forms of cognition, some within organizations, and some in networks and markets. Anthropology provides tools for understanding cultures of thought within organizations and groups. Computer science supplies rigorous methods for understanding logics of processing, pattern recognition, and learning. Psychology offers ways of understanding group dynamics, and how to enhance or impair individual intelligence. Philosophy provides tools for thinking about thinking. My hope is that over the coming decades, different syntheses will emerge, some using the concepts set out in this book, to inform an emerging field of study that can analyze systems, patterns of thought, and problem solving.

Experiment and Research

Much of the daily work of an emerging discipline is, and should be, descriptive and analytic: observing collective intelligence "in the wild" to see how together machines, organizations and groups think, and why some think more successfully than others. But to advance it will also require experiment. A lot of research has been done in the last decade on collective intelligence in computing—understanding the cultures, motivations, and practical methods of open-source software—and there is vastly more scope for empirical study, such as how platforms do or don't help with coordination, creativity, or action; how rewards work for collaboration and the challenges of sustaining sharing in projects like Wikipedia; and the practical lessons to be learned from labs, accelerators, or open innovation tools. There has also been a lot of research on the psychology of groups, though some shares the weakness of much recent psychology: an overdependence on experiments with small groups of North American undergraduates who, despite plenty of evidence to the contrary, are taken to stand in for the whole of humanity.

So we should hope for energetic experiment, new disciplines, and subdisciplines emerging to map the territory of this landscape. Some of the projects I have been involved in over the last few years point in this direction.[11] Hundreds of others are under way in science, medicine, and both academia and civil society, providing a rich trove for researchers to study.

One route for the study of collective intelligence could be a series of axioms and knowledge that could be deduced. This was the goal of modern economics, physics, and some other fields. I expect that this field will advance in a rather different way, though, using the same underlying concepts, but mapping a complex landscape with detailed description of patterns, more along the lines of chemistry than physics. This is how we will understand which assemblies work, and for which tasks. It will be a more laborious task than the discovery of a handful of generative axioms. But it will tell us more about both our future and past.

SUMMARY OF THE ARGUMENT

CollecTIVE INTELLIGENCE IS THE CAPACITY of groups to make good decisions—to choose what to do, and who to do it with—through a combination of human and machine capabilities. The ways intelligence is organized are *largely fractal in nature* with similar patterns occurring on multiple scales, from groups of friends to organizations and whole societies.

On each scale, collective intelligence depends on *functional capabilities*: distinct abilities to observe, analyze, remember, create, empathize, and judge—each of which can be enhanced by technologies, and each of which also has a cost.

These are then supported by *infrastructures* that make collective intelligence easier: common standards and rules, physical objects that embody intelligence, institutions that can concentrate the resources needed for the hard work of thought, and looser networks and societies of mind. Some of the most important recent infrastructures are hybrid combinations of machine and human intelligence, orchestrated on large scales.

Making the most of these capabilities and infrastructures depends on *models of organization* that assemble capabilities and infrastructures in ways that allow for continuous learning. The most successful ones have five characteristics: they create autonomous knowledge and informational commons, achieve an appropriate balance between their functional capabilities, achieve focus, orchestrate systematic reflection, and integrate for action. Powerful tendencies in organizations and societies—including conflicting interests—push in opposite directions to each of these, which is why they are rare.

The everyday *processes of intelligence* then operate at multiple levels that link together in a hierarchy: a first loop using existing models to process data, a second loop of learning that generates new categories and relationships, and a third loop that creates new ways of thinking. These can be combined in triggered hierarchies.

Groups and organizations think well when they have all these in place, with a balance between capabilities, effective infrastructures, systematic ways of managing the three loops, and a willingness to devote resources to the hard work of structured thought, to tap into a bigger mind beyond their own boundaries and remain self-conscious about methods.

But most important fields of human activity lack crucial elements—and so end up much more collectively stupid than they could be. The spread of the Internet along with ubiquitous tools for analysis, search, and memory have greatly enhanced the world's capacity to think. Many more resources are devoted to collective intelligence in competitive fields than cooperative ones, however, and the world suffers from a huge *misallocation of brainpower* as well as machine intelligence.

The successful examples of collective intelligence are best understood as *assemblies* of multiple elements. Discovering which assemblies work best requires continuous shuffling of the elements, since capabilities, infrastructures, and organizational models have to coevolve with environments. Yet some of the most important fields—including politics, the university, and finance—lack this capacity for iterative shuffling, and so become locked into configurations that keep them less effective than they should be.

At a global level, there is a *need for new assemblies* that can marshal global collective intelligence for global tasks, from addressing climate change to avoiding pandemics, solving problems of unemployment to the challenges of aging. It is possible to imagine what these could look like—building on recent initiatives in medicine and the environment that try to observe, model, predict, and act. Creating such tools on a scale, and with capabilities proportionate to the challenges, and nurturing people with skills in "intelligence design" will be one of the great tasks facing the twenty-first century.

NOTES

Introduction: Collective Intelligence as a Grand Challenge

1. There is a vast literature on individual intelligence—covering whether it is one thing or many, how much it is inherited or cultivated, and its use of heuristics and shortcuts—by many leading thinkers such as Robert Sternberg. There is also a huge literature on artificial intelligence (such as Nick Bostrom's recent and excellent book on superintelligence). But there are few concepts, theories, or data on intelligence as something collective (though at the end of the book, I offer a survey of some of the relevant literatures).

2. A number predicted to rise to twenty billion by 2020.

3. The company DeepMind describes its mission as being to solve intelligence, and having done that, then solve everything else. I don't make anything like so bold a claim. But better understanding collective intelligence in all its forms potentially has wide applicability.

4. Over the last thirty years, almost all the most influential writers about networks and the Internet have contributed to a lopsided view of this question, combining important truths with what in retrospect look more like fairy tales. Indeed their claim that technology would inevitably make the world freer, more democratic, and more equal arguably diverted many people from the hard work necessary to actually achieve that goal.

5. A parallel and similarly uneven distribution can be seen in the skills surrounding new tools for collective intelligence—from data and machine learning to online collaboration—which are highly concentrated in elite groups, primarily in a handful of cities around the world. Twenty-first-century policy makers will have to find much better answers to the questions of how to direct human and machine brainpower to where it's most needed, and how to widen access to the best tools for amplifying human intelligence.

Chapter 1: The Paradox of a Smart World

1. William Gibson, *Neuromancer* (New York: Berkley Publishing Group, 1989), 128.

2. Sherry Turkle, "Artificial Intelligence at Fifty: From Building Intelligence to Nurturing Sociabilities" (paper presented at the Dartmouth Artificial Intelligence Conference, Hanover, NH, July 15, 2006), accessed April 13, 2017, http://www.mit.edu/~sturkle/ai@50.html.

3. Some people in finance argue that their decisions were not stupid: because others (mainly governments) essentially guaranteed the risks, investors, and banks were wise to act recklessly.

4. The *Journal of Medical Internet Research* is a good source, showing a steady improvement in the reliability of guidance on social media, but also major cultural differences around the world.

5. Maria Frellick, "Medical Error Is the Third Leading Cause of Death in the US," Medscape, May 3, 2016, accessed April 13, 2017, http://www.medscape.com/viewarticle /862832.

6. Evgeny Morozov, *To Save Everything, Click Here* (New York: Public Affairs, 2013); Nicholas Carr, *The Glass Cage: Automation and Us* (New York: W. W. Norton, 2014).

7. Some past eras might have been better at handling the really big questions of collective intelligence. Think, for example, of the wave of nation building that took place after World War II, remaking Japan, Germany, Korea, and China, or the design of global institutions in the same period from Bretton Woods to the United Nations. At the very least, we shouldn't be too self-satisfied.

Chapter 2: The Nature of Collective Intelligence in Theory and Practice

1. Islamic scholars led much of this thinking. See Herbert Davidson, *Alfarabi, Avicenna, and Averroes, on Intellect: Their Cosmologies, Theories of the Active Intellect, and Theories of Human Intellect* (Oxford: Oxford University Press, 1992).

2. For an excellent overview of how to avoid being trapped by these metaphors, and how to assess intelligence without falling into the trap of excess anthropocentrism, see José Hernández-Orallo, *The Measure of All Minds: Evaluating Natural and Artificial Intelligence* (Cambridge: Cambridge University Press, 2016).

3. There are innumerable definitions of intelligence. I like one of Robert J. Sternberg's early ones: "Intelligent behaviour involves adapting to your environment, changing your environment, or selecting a better environment." Others can be found in his book *Wisdom, Intelligence, and Creativity Synthesized* (Cambridge: Cambridge University Press, 2007). For a different set of definitions, see Thomas W. Malone and Michael S. Bernstein, eds., *Handbook of Collective Intelligence* (Cambridge, MA: MIT Press, 2015). Theorists of IQ define it as the ability to perform a range of cognitive tasks; the *Encyclopaedia Britannica* defines it as an ability to adapt to the environment, while the psychologist Howard Gardner described it as a set of abilities to solve problems that are valued in particular cultures. See Howard Gardner, *Frames of Mind: The Theory of Multiple Intelligences* (New York: Basic Books, 1983).

4. Thucydides, *The History of the Pelopponesian War*, 3.20, accessed May 31, 2017, http://www.gutenberg.org/ebooks/7142.

5. To adopt the phrase used in Daniel Dennett, *From Bacteria to Bach and Back* (New York: W. W. Norton, 2017).

6. Decision theorists have tried to define a good decision as one that is logically consistent and in which the chosen means are consistent with the desired ends. That is a demanding standard; it's even harder to justify the ends being chosen and why they are superior to all the possible alternatives.

7. A parallel debate is under way in law about the potential for artificial intelligence to generate "self-driving" laws that adapt automatically to changes in the environment. See Anthony J. Casey and Anthony Niblett, "Self-Driving Laws," *University of Toronto Law Journal* 66, no. 4 (2016), accessed April 13, 2017, http://www.utpjournals.press/doi/abs /10.3138/UTLJ.4006.

8. This vision of unrealized possibility draws on the work of Roberto Mangabeira Unger, particularly his book *The Self Awakened: Pragmatism Unbound* (Cambridge, MA: Harvard University Press, 2007) and his essay on social innovation in Alex Nicholls, Julie

Simon, and Madeleine Gabriel, eds., *New Frontiers of Social Innovation* (Houndmills, UK: Palgrave Macmillan, 2015).

9. Planet Labs, accessed April 13, 2017, https://www.planet.com.

10. These tools could achieve a degree of economic transparency that will be much harder to distort or game (though, like any metric, they may in time be gamed, with fake trucks, just as armies throughout history created fake camps, fires, and noises to confuse the enemy).

11. While most of these build on the Internet's capacity to share information at low cost, another family of innovations is growing up around the blockchain's capacity to move, store, or protect assets that are valuable, such as money, votes in elections, titles, deeds, or works of art, allowing trust between strangers and orchestrating the world's memory in novel ways. Blockchain may turn out to be a transitional technology superseded by other, more flexible distributed ledgers. But the principle of general transparency and open verification along with the use of automated tools to ensure that data have not been tampered with and are used in accordance with agreed-on policies is likely to become more general. There are likely to be many kinds of "distributed ledger" in the near future, where data can be added but not taken away. In principle, inputs cannot subsequently be tampered with, and there is no need to rely on a single store or primary copy. Smart contracts with "executable code," which automatically act in particular ways (for instance, releasing money to a specified account when particular actions have been done) are a good example. It remains unclear, though, how these technologies will evolve, how much they will depend on intermediaries, whether they will replicate the Bitcoin model of depending on many computers or will rely on a few, how they will avoid the risks of being hacked or taken over, or indeed whether they really will enhance the world's collective intelligence by organizing memory in more efficient ways.

12. The Intelligence Advanced Research Projects Activity in the United States has funded the Machine Intelligence from Cortical Networks program to observe animals learning tasks and map the connections used to build better machine learning.

13. Iyad Rahwan, Sohan Dsouza, Alex Rutherford, Victor Naroditskiy, James Mc-Inerney, Matteo Venanzi, Nicholas R. Jennings, and Manuel Cebrian, "Global Manhunt Pushes the Limits of Social Mobilization," *Computer* 46 (2013): 68–75, doi:10.1109/mc.2012.295.

14. M. Mitchell Waldrop, *The Dream Machine: J.C.R. Licklider and the Revolution That Made Computing Personal* (New York: Viking Penguin, 2001), dust jacket.

15. Tim Berners-Lee and Mark Fischetti, *Weaving the Web: The Original Design and Ultimate Destiny of the World Wide Web* (New York: HarperCollins, 2000), 172.

16. Francis Heylighen, "From Human Computation to the Global Brain: The Self-Organization of Distributed Intelligence," *Handbook of Human Computation* (New York: Springer, 2013), 897–909.

17. The move to an open API was partly prompted Paul Rademacher, who independently created a "mash up" of house prices for which he reverse engineered the Google Maps code.

18. It originated in a corner of the European Organization for Nuclear Research center, but was given away as open code, unlike its rival, Gopher, a text-based information-linking system created at the University of Minnesota, which was turned into a source of revenue—an emblematic case of greed leading to self-destruction.

19. Daniela Retelny, Sébastien Robaszkiewicz, Alexandra To, Walter Lasecki, Jay Petel, Negar Rahmati, Tulsee Doshi, Melissa Valentine, and Michael S. Bernstein,

"Expert Crowdsourcing with Flash Teams," accessed April 14, 2017, http://hci.stanford .edu/publications/2014/flashteams/flashteams-uist2014.pdf. Hybrids are common in business. The company Palantir, which dominates US security and intelligence, is famous for its data analytics. It deploys its engineers to work closely with intelligence analysts so that they can better understand their needs, building large databases in the background, but with a constantly adapted and flexible front end, so that analysts can write arbitrary queries with a reasonable chance of getting a meaningful response.

20. WikiHouse Foundation, accessed April 14, 2017, https://www.wikihouse.cc/.

21. Health has many equivalents. The C-Path system is one case. Breast cancer prognosis used to depend on doctors identifying three specific features through a microscope to estimate the likely survival rate for the patient. The C-Path program measures many more features (nearly seven thousand) in the breast cancer and surrounding tissue, and performs much better than humans in analyzing and evaluating images. It has spotted unknown features that turned out to be better predictors, which then allows for better screening of precancerous tissues. These then allow it to improve the underlying model. But it continually incorporates feedback from the doctors using the system.

22. Jon Kleinberg, Jens Ludwig, and Sendhil Mullainathan, "A Guide to Solving Social Problems with Machine Learning," *Harvard Business Review*, December 8, 2016, accessed April 14, 2017, https://hbr.org/2016/12/a-guide-to-solving-social-problems-with -machine-learning.

23. For an excellent account of the economics of algorithmic error, see Juan Mateos Garcia's blog, accessed May 31, 2017, http://www.nesta.org.uk/blog/err-algorithm-algorithmic -fallibility-and-economic-organisation.

24. All also run up against the constraints of computing power. As more variables are included, the complexity rises exponentially. So it's the interaction of professionals and algorithm that is key: the algorithm challenges the professionals, pointing them to examples they hadn't thought of. But their experience also helps guide the algorithm.

25. Michael Nielsen, *Reinventing Discovery: The New Era of Networked Science* (Princeton, NJ: Princeton University Press, 2012).

26. In all this work two ironies stand out. One is that the more we try to replicate human with machine intelligence, the more we have to ask more fundamental questions of what intelligence is. The other is that in order to make machines think more like us, we have to think more like machines, in algorithmic, Bayesian ways (while in myriad other ways people are also having to become more robot-like in their daily lives in order to coexist with automated machines).

27. "Limits of Social Learning," MIT Media Lab, accessed April 16, 2017, http:// socialphysics.media.mit.edu/blog/2015/8/4/limits-of-social-learning.

28. We can roughly summarize some of what is known about what works—and the advantages and disadvantages of widening the pool of people, or machines, involved in solving a problem or completing a task:

- Task definition: Much depends on the nature of the task at hand. How fast does it need to be done, what resources are available, and how novel is it (are there ready solutions at hand)?
- Quantity: The optimum number of brains and machine intelligences involved in a task then comes up against trade-offs, such as between the simplicity of selection criteria and the quantity. Simple selection criteria (which make it easy to see if a

problem has been solved or not) make it easier to open much wider. With fuzzy or ambiguous selection criteria, more people involved may just mean more noise and confusion.

- Quality: Parallel considerations relate to the quality of the brains or machine intelligence being mobilized—their knowledge, experience, and capability. Some specialized tasks may be best performed by a small number of highly skilled people or machines (though here too there will be trade-offs: knowledgeable people may find it harder to imagine new answers).
- Organization: Finally, much depends on how well organized the various intelligent resources are. This will include a division of labor, the sequencing of tasks, and coordination. Again there will be trade-offs. Effective organizations may struggle if the environment changes or the task is wholly novel.

29. In the words of one of the pioneers of web science, "Once upon a time 'machines' were programmed by programmers and used by users. The success of the Web has changed this relationship: we now see configurations of people interacting with content and with each other, typified by social websites. Rather than drawing a line through such Web-based systems to separate the human and digital parts (as computer science has traditionally done), we can now draw a line around them and treat each such compound as a social machine—a machine in which the two aspects are seamlessly interwoven." Nigel R. Shadbolt, Daniel A. Smith, Elena Simperl, Max Van Kleek, Yang Yang, and Wendy Hall, "Towards a Classification Framework for Social Machines" (paper presented at the twenty-second International Conference on World Wide Web, Rio de Janeiro, Brazil, May 13–17, 2013), accessed April 16, 2017, http://sociam.org/www2013/papers/socm2013_submission_9.pdf.

30. "Building a 'Google Earth' of Cancer," National Physical Laboratory, accessed April 18, 2017, http://www.npl.co.uk/grandchallenge/?utm_source=weeklybulletin&utm_medium=email&utm_campaign=iss290; MetaSub project, Weill Cornell Medical College, New York, accessed April 18, 2017, http://metasub.org/.

31. AIME for Life, accessed April 18, 2017, http://aime.life/.

32. The examples mentioned above pull together some research funding, philanthropic funding, and funding from governments when they help them solve problems. Few have a stable long-term funding base suitable for what are in effect essential parts of the world's nervous system.

Chapter 3: The Functional Elements of Collective Intelligence

1. For a range of recent overviews, see Roberto Colom, Sherif Karama, Rex E. Jung, and Richard J. Haier, "Human Intelligence and Brain Networks," *Dialogues in Clinical Neuroscience* 12, no. 4 (2010): 489–501, accessed April 18, 2017, https://www.ncbi.nlm.nih.gov/pmc/articles/PMC3181994/; Linda S. Gottfredson, "Mainstream Science on Intelligence: An Editorial with 52 Signatories, History, and Bibliography," *Intelligence* 24, no. 1 (1997): 13–23, accessed April 18, 2017, http://www.intelligence.martinsewell.com/Gottfredson1997.pdf; Earl Hunt, *Human Intelligence* (Cambridge: Cambridge University Press, 2011).

2. The same is true of our brains. The eighty-six billion neurons packed into fourteen hundred grams that help humans think require 25 percent of our energy, compared to 10 percent for other vertebrates.

3. I deliberately avoid here the many attempts in psychology to define the key elements of individual intelligence, such as Howard Gardner's eight or nine elements of multiple intelligence, and Dan Sperber's suggestion of specific cognitive modules like snake detection and facial recognition. For a description, see Dan Sperber and Lawrence A. Hirschfeld, "The Cognitive Foundations of Cultural Stability and Diversity," *Trends in Cognitive Sciences* 8, no. 1 (2004): 40–46. There continues to be little agreement on what these all are, how many there are, or what boundaries there are between them. Steven Mithen argues that in the early period of human development, there were a series of separate facilities of cognition—tools, animals, and a social intelligence with other people—but that these were not integrated. See Steven J. Mithen, *The Prehistory of the Mind: The Cognitive Origins of Art, Religion, and Science* (London: Thames and Hudson, 1996).

4. Psychology and neuroscience have been through endless debates about whether individual intelligence is something general or made up of distinct elements (a weakness of most current theories is that they are inherently hard to falsify). Some examples include theories of multiple intelligences, Piagetian theories, Luria's PASS theories, Sternberg's various theories such as the triarchic theory, and many others.

5. For a good overview of the broader field of understanding causes, see Judea Pearl, "Causal Inference in Statistics: An Overview," *Statistics Surveys* 3 (2009): 96–146, doi: 10.1214/09-ss057.

6. For this example, see Michael Tomasello, *A Natural History of Human Thinking* (Cambridge, MA: Harvard University Press, 2014).

7. This was one of the messages of Aaron Antonovsky's work on health and resilience: having a "sense of coherence"—a meaningful account of your place in the world—is valuable for both physical and mental health, and often that involves feeling useful. Aaron Antonovsky, *Unraveling the Mystery of Health: How People Manage Stress and Stay Well* (San Francisco: Jossey-Bass, 1987).

8. Few of our models are our own. Ludwig Fleck, the great Polish epidemiologist, wrote in his book *The Genesis and Development of a Scientific Fact* (Chicago: University of Chicago Press, 1979) that "knowledge is the paramount social creation," and that the prevailing thought style in your social group "exerts an absolutely compulsive force upon [one's] thinking . . . with which it is not possible to be at variance." A few do vary and rebel, but people are highly suggestible and extraordinarily prone to copy others, and this makes collaboration easy (and much easier for humans than for other apes). Our ability to become a collective, to identify with a larger group and submerge ourselves within a bigger whole, is both one of the great aids to collective intelligence and also, as I will show later, one of the great hindrances, since groups tend to be defined by what they ignore and what they forget, as much as by what they know.

9. Nelson Cowan, "The Magical Number 4 in Short-Term Memory: A Reconsideration of Mental Storage Capacity," *Behavioral and Brain Sciences* 24, no. 1 (2001): 87–114.

10. Cognitive science makes various distinctions to understand human memory, such as the distinction between declarative memory (Who fought World War II?) and procedural memory (How do I ride a bike?). Parallel distinctions are likely needed for collective memory. For an interesting recent piece of research on the latter, see Ruth García-Gavilanes, Anders Mollgaard, Milena Tsvetkova, and Taha Yasseri, "The Memory Remains: Understanding Collective Memory in the Digital Age," *Science Advances* 3, no. 4 (April 2017): e1602368, doi: 10.1126/sciadv.1602368.

11. Everledger and Provenance are two start-ups attempting each of these tasks.

12. Leon A. Gatys, Alexander S. Ecker, and Matthias Bethge, "A Neural Algorithm of Artistic Style," August 26, 2015, accessed April 18, 2017, https://arxiv.org/abs/1508 .06576.

13. For a summary of some of the debate in Japan about "wisdom computing," see Iwano Kazuo and Motegi Tsuyoshi, "Wisdom Computing· Toward Creative Collaboration between Humans and Machines," *Joho Kanri: Journal of Information Processing and Management* 58, no. 7 (2015): 515–24, accessed April 18, 2017, https://www.jstage.jst.go .jp/article/johokanri/58/7/58_515/_pdf.

14. Much of this recent knowledge warns against seeing too many similarities between human thought and computer thought. As Robert Epstein put it, contesting much of the conventional wisdom of his field, "We are not born with: information, data, rules, software, knowledge, lexicons, representations, algorithms, programs, models, memories, images, processors, subroutines, encoders, decoders, symbols, or buffers—design elements that allow digital computers to behave somewhat intelligently. Not only are we not born with such things, we also don't develop them—ever. We don't store words or the rules that tell us how to manipulate them. We don't create representations of visual stimuli, store them in a short-term memory buffer and then transfer the representation into a long-term memory device." Instead, our brains are better understood as capabilities, good at responding to environments, stimuli, and of course other people. Robert Epstein, "The Empty Brain," Aeon, May 18, 2016, accessed April 18, 2017, https://aeon.co/essays/your-brain-does-not -process-information-and-it-is-not-a-computer.

15. The philosophical debate about whether an intelligence serves only itself is unresolved. It's possible that this leads to an infinite regress and incoherence. My view is that it's only when intelligence serves something else—a body, life, or thing grounded in time and space—that it can make sense of itself.

16. Rogers Hollingworth, "High Cognitive Complexity and the Making of Major Scientific Discoveries," in *Knowledge, Communication, and Creativity*, ed. Arnaud Sales and Marcel Fournier (London: Sage, 2007), 149.

17. Much of the hard work of statistics and machine learning involves trying to reduce this dimensionality, using tools like principal component analysis.

Chapter 4: The Infrastructures That Support Collective Intelligence

1. Simon Winchester, *The Professor and the Madman: A Tale of Murder, Insanity, and the Making of the Oxford English Dictionary* (New York: Harper Perennial, 2005), 106. My father and one cousin both worked on the *OED*, working on revisions for letter *A*.

2. Information science uses the word ontology in different ways from philosophy to describe the rules that formalize how information is organized.

3. Jessica Seddon and Ramesh Srivinasan, "Information and Ontologies: Challenges in Scaling Knowledge for Development," *Journal of the Association for Information Science and Technology* 65, no. 6 (2014): 1124–33.

4. On how the question "What works?" needs a series of supplementary questions, see Geoff Mulgan, "The Six Ws: A Formula for What Works," accessed April 20, 2017, http:// www.nesta.org.uk/blog/six-ws-formula-what-works.

5. See, among many others, Edwin Hutchins, *Cognition in the Wild* (Cambridge, MA: MIT Press, 1995); Randall D. Beer, *Intelligence as Adaptive Behavior: An Experiment in Computational Neuroethology* (New York: Academic Press, 1989).

6. Alex Bell, Raj Chetty, Xavier Jaravel, Neviana Petkova, and John Van Reenen, "The Lifecycle of Inventors," June 13, 2016, accessed April 20, 2017, https://www.rajchetty.com /chettyfiles/lifecycle_inventors.pdf.

7. For what remains probably the best account of how cities nurtured new knowledge at different stages of history, see Peter Hall, *Cities in Civilization* (New York: Pantheon, 1998).

8. Sandro Mendonca, "The Evolution of New Combinations: Drivers of British Maritime Engineering Competitiveness during the Nineteenth Century" (PhD diss., University of Sussex, 2012); Sidney Pollard, *Britain's Prime and Britain's Decline: The British Economy, 1870–1914* (London: Edward Arnold, 1990), 189.

9. Mott Greene, "The Demise of the Lone Author," *Nature* 450 (2007): 1165.

10. Stefan Wuchty, Benjamin F. Jones, and Brian Uzzi, "The Increasing Dominance of Teams in Production of Knowledge," *Science* 316 (2007): 1036–39.

11. Karl R. Popper, *The Open Society and Its Enemies* (Princeton, NJ: Princeton University Press, 1945), 82.

Chapter 5: The Organizing Principles of Collective Intelligence

1. Karen Eisenstadt, "High Reliability Organizations Meet High Velocity Environments," in *New Challenges to Understanding Organizations*, ed. Karlene H. Roberts (New York: Macmillan, 1993), 132.

2. This section draws on Geoff Mulgan, *The Locust and the Bee: Predators and Creators in Capitalism's Future* (Princeton, NJ: Princeton University Press, 2013).

3. Such as AMEE (https://www.amee.com/), which began as an open-data approach to carbon before morphing into a set of tools for supply chains.

4. Chou En-lai famously commented on the impact of the French Revolution that it was too soon to say (though he appears to have been referring to the revolution of 1968 rather than 1789, making the comment less elliptically profound that it appeared). The IPCC, like both 1789 and 1968, may have far more long-term impact than is apparent in the immediate aftermath.

5. What could have been done differently? For all its ambition, the IPCC didn't attempt to create a truly distributed intelligence, dependent on hundreds of millions of citizens' views and voices as well as media, NGOs, and businesses. Its model was much better at analysis than prescription or creativity. And it lacked genuine reflexiveness—the capacity to critique itself. Its biggest challenge was the sheer scale of the problem or the network of linked problems—all feeding into carbon dioxide levels and climate change, but so different in nature, such as how to legitimate new laws or taxes, how to change traffic or air, how to deal with buildings regulations or trading schemes, and how to change everyday behavior. Even more seriously, the IPCC lacked sufficient methods for synthesis—for integrating political, economic, ecological, and other factors. Of course, no other institutions have strong methods of this kind either, and so we end up leaving the task of synthesis to harried politicians, trying to weigh multiple factors in an ad hoc way.

6. I use the word *recursive* in its original sense, as in a circular looping back to reconsider something. In software, the word has taken on a different meaning, with a recursive structure containing smaller versions of itself—potentially to many degrees.

7. These ideas draw on Chris Argyris's work on double-looped learning, and Donald Schon's work on reflective practice.

8. I am adapting the famous comment by Oliver Wendell Holmes that he wouldn't give a fig for the simplicity this side of complexity, but instead would give his life for the simplicity on the other side of complexity.

Chapter 6: Learning Loops

1. This framework draws on yet extends the famous distinction made by Argyris and Schon between single-loop learning, which learns from new facts, but doesn't question the goal or logic being followed, and double-loop learning that asks the bigger questions. See Chris Argyris and Donald A. Schon, *Organizational Learning* (Reading, MA: Addison-Wesley, 1978). For a similar framework, see James March, "Exploration and Exploitation in Organizational Learning," *Organization Science* 2, no. 1 (1991): 71–87.

2. A modern equivalent of the Turing test would probably want to assess abilities to reason at these three levels—rather than just the ability to appear like a human. For the third level it might ask whether the machine intelligence can generate a novel *fragestellung*, the German word meaning a worldview, but that is literally the posing of a question that makes it possible to see things and ask in new ways.

3. I like this comment from Steven Pinker on whether robots will produce literature: "Intelligent systems often best reason by experiment, real or simulated: they set up a situation whose outcome they cannot predict beforehand, let it unfold according to fixed causal laws, observe the results and file away a generalization about how what becomes of such entities in such situations. Fiction, then, would be a kind of thought experiment, in which agents are allowed to play out plausible interactions in a more-or-less lawful virtual world and an audience can take mental notes of the results. Human social life would be a ripe domain for this experiment-driven learning because the combinatorial possibilities in which their goals may coincide and conflict (cooperating or defecting in prisoner's dilemmas, seeking long-term or short-term mating opportunities, apportioning resources among offspring) are so staggeringly vast as to preclude strategies for success in life being either built-in innately or learnable from one's own limited personal experience." Steven Pinker, "Toward a Consilient Study of Literature," *Philosophy and Literature* 31, no. 1 (2007): 172.

Chapter 7: Cognitive Economics and Triggered Hierarchies

1. Gautam Ahuja, Giuseppe Soda, and Akbar Zaheer, "The Genesis and Dynamics of Organizational Networks," *Organization Science* 23 (2012): 434–48.

2. More than a century ago, the sociologist Gabriel Tarde proposed the idea of monads as a way of understanding this dual character of human organization, which combines self-organization and being organized. He recommended breaking away from what he saw as the false distinction between the individual and society. The individual is both apart and a part, defined by difference, but also by connections.

3. Delving further into dimensionality, we could also distinguish the distribution of payoffs (positive and negative) of the action to be taken for each participant, or the dynamics of when costs are entailed and different streams of benefits are created.

4. See Juan Mateos Garcia, "To Err Is Algorithm: Algorithmic Fallibility and Economic Organisation," May 10, 2017, accessed May 31, 2017, http://www.nesta.org.uk/blog/err-algorithm-algorithmic-fallibility-and-economic-organisation.

5. See various chapters in Don Ambrose and Robert J. Sternberg, *Creative Intelligence in the 21st Century: Grappling with Enormous Problems and Huge Opportunities* (Rotterdam: Sense Publishers, 2016).

6. Here's why I don't use the term *meme* in this book. Although much of collective intelligence looks at first glance to involve the spread of memes, the word is misleading and adds nothing to the word *idea*. Indeed, by appearing to show ideas as similar to genes, it risks creating confusion. Unlike genes, memes are not reproduced with near-precise accuracy but instead distort and decay. They are not produced by random mutations but rather by nonrandom ones (including the influence of theories of mind, as creators of ideas try to imagine how others will receive them). They are not selected in the same way by fitness, and instead bad ideas can spread as easily as good ones if they have the right properties of attraction.

7. Robert L. Helmreich and H. Clayton Foushee, "Why Crew Resource Management: Empirical and Theoretical Bases of Human Factors Training in Aviation," in *Cockpit Resource Management*, ed. Earl L. Wiener, Barbara G. Kanki, and Robert L. Helmreich (San Diego, CA: Academic Press, 1993), 3–46.

8. The oddly titled "Scientific Community Metaphor" methods and associated software like Ether were explicitly designed to support a community as a collective intelligence, able to show the links between scientific ideas. See William A. Kornfeld and Carl E. Hewitt, "The Scientific Community Metaphor," *IEEE Transactions on Systems, Man, and Cybernetics* 11, no. 1 (1981): 24–33.

9. In *The Nerves of Government: Models of Political Communication and Control* (New York: Free Press, 1963), Karl Deutsch wrote that power was the ability to not have to learn.

10. See, for example, George A. Miller, "The Cognitive Revolution: A Historical Perspective," *Trends in Cognitive Sciences* 7, no. 3 (2003): 141–44.

11. Rodney A. Brooks, "Intelligence without Representation," *Artificial Intelligence* 47 (1991): 139–59.

12. What Hubert Dreyfus calls "smooth coping," the ability to act without thinking (and without representations) in response to an environment, echoes Gary Klein's famous work on firefighters and their ability to learn rapid heuristics that they cannot easily explain.

13. Michael Polanyi, *The Tacit Dimension* (New York: Doubleday and Company, 1966).

Chapter 8: The Autonomy of Intelligence

1. Karl Duncker, *Zur Psychologie des produktiven Denkens* (Berlin: Springer, 1935).

2. This was the slogan we devised at Nesta for the Alliance for Useful Evidence.

3. Alan M. Leslie, "Pretense and Representation: The Origins of Theory of Mind," *Psychological Review* 94, no. 4 (1987): 412–26.

4. This was originally called a Universal Resource Identifier.

Chapter 9: The Collective in Collective Intelligence

1. Lucy Kellaway, "I Have Fallen into Recession's Web of Fear," *Financial Times*, February 1, 2009, accessed April 23, 2017, https://www.ft.com/content/f55ee4ca-996e-45b4 -a2f9-07f2b90d3745.

2. See Abraham Sesshu Roth, "Shared Agency," December 13, 2010, accessed April 23, 2017, https://plato.stanford.edu/entries/shared-agency/; Deborah Perron Tollefsen, "Groups as Agents," *Polity*, May 2015, accessed April 23, 2017, http://eu.wiley.com /WileyCDA/WileyTitle/productCd-0745684831.html.

3. David Chalmers adopted a similar distinction in his book *The Conscious Mind* (New York: Oxford University Press, 1996) and described the remaining problem of consciousness as the hard issue.

4. As a matter of law, if a group is not incorporated, then it consists only of individuals that the law defines as members of the group; laws can be passed that give individual members of those defined groups specific rights or responsibilities. If a group is incorporated, its corporation has an individual legal identity and can be called to account for its actions. And the individuals who held offices in the corporation can also be found individually guilty or innocent of the crimes of the corporation.

5. Giulio Tononi and Christof Koch, "Consciousness: Here, There, and Everywhere?" *Philosophical Transactions of the Royal Society B*, 370, no. 1668 (2015): 20140167.

6. For relevant literature, see Mattia Gallotti and Chris Frith, "Social Cognition in the We-Mode," *Trends in Cognitive Sciences* 17, no. 4 (2013): 160–65; Julian Kiverstein, ed., *The Routledge Handbook of Philosophy of the Social Mind* (London: Routledge, 2017); Michael P. Letsky, Norman W. Warner, Stephen M. Fiore, and C.A.P. Smith, eds., *Macrocognition in Teams: Theories and Methodologies* (Aldershot, UK: Ashgate Publishing, 2008).

7. Wilfrid Sellars called this the "we mode." See Wilfrid Sellars, *Science and Metaphysics* (London: Routledge and Kegan Paul, 1968). The concept has been theorized about and popularized by the Finnish philosopher Raimo Tuomela, and then developed by more recent scholars such as Mattia Gallotti.

8. Michael Tomasello, *Origins of Human Communication* (Cambridge, MA: MIT Press, 2008); Henrike Moll and Michael Tomasello, "Cooperation and Human Cognition: The Vygotskian Intelligence Hypothesis," *Philosophical Transactions of the Royal Society B: Biological Sciences* 362, no. 1480 (2007): 639–48.

9. For an interesting approach to game theory, see Michael Bacharach, *Beyond Individual Choice: Teams and Frames in Game Theory* (Princeton, NJ: Princeton University Press, 2006).

10. Cooperation and empathy can support each other, but don't necessarily come together. I can cooperate with others without in any way empathizing with them. And I can empathize with an enemy.

11. Martin A. Nowak and Karl Sigmund, "Evolution of Indirect Reciprocity," *Nature* 437, no. 7063 (2005): 1291–98.

12. Karl Friston and Christopher Frith, "A Duet for One," *Consciousness and Cognition*, accessed April 23, 2017, http://www.fil.ion.ucl.ac.uk/~karl/A%20Duet%20for%20one.pdf.

13. The theory of distributed cognition uses the measuring tape that embodied a system of comparable measurements and ease of use as an example. See Edwin Hutchins, *Cognition in the Wild* (Cambridge, MA: MIT Press, 1995).

14. Karl E. Weick, "The Collapse of Sensemaking in Organizations: The Mann Gulch Disaster," *Administrative Science Quarterly* 38 (1993): 628–52.

15. Andy Clark, "Whatever Next? Predictive Brains, Situated Agents, and the Future of Cognitive Science," *Behavioral and Brain Sciences* 36, no. 3 (June 2013): 181–204, accessed April 23, 2017, https://www.cambridge.org/core/journals/behavioral-and-brain-sciences /article/div-classtitlewhatever-next-predictive-brains-situated-agents-and-the-future-of -cognitive-sciencediv/33542C736E17E3D1D44E8D03BE5F4CD9.

16. Garold Stasser and Beth Dietz-Uhler, "Collective Choice, Judgment, and Problem Solving," *Blackwell Handbook of Social Psychology: Group Processes* 3 (2001): 31–55; Janet B. Ruscher and Elliott D. Hammer, "The Development of Shared Stereotypic Impressions in Conversation: An Emerging Model, Methods, and Extensions to Cross-Group Settings," *Journal of Language and Social Psychology* 25, no. 3 (2006): 221–43. Some of the classic literature includes Barry E. Collins and H. Guetzkow, *A Social Psychology of Group Processes for Decision Making* (New York: Wiley, 1964); J. H. Davis, *Group Performance* (New York: Addison-Wesley, 1969); Ivan D. Steiner, *Group Process and Productivity* (New York: Academic Press, 1972).

17. Mark Warr, *Companions in Crime: The Social Aspects of Criminal Conduct* (Cambridge: Cambridge University Press, 2002).

18. Simon Hartley, *Stronger Together: How Great Teams Work* (London: Piatkus, 2015).

19. For a description of Google's research on teams, called Project Aristotle, see Charles Duhigg, *Smarter Faster Better: The Secrets of Being Productive* (New York: Random House, 2016).

20. There is a semiscience available for thinking about how different kinds of voting schemes, with varied weightings or sequencings, can be used in groups to avoid foolish or precipitous decisions. We also know that within teams, members are more likely to share knowledge and information that is already known, or that feels comfortable, than information that is unsettling or not already shared with others. This is a counterintuitive finding from research, since you might expect teams to be best at sharing information that isn't already shared. It's one of the reasons why good teams make extra efforts to encourage more brutal honesty and share what's useful rather than what's convenient.

21. Jon Elster, *Explaining Social Behavior: More Nuts and Bolts for the Social Sciences* (Cambridge: Cambridge University Press, 2015), 368

22. Similar patterns could be found elsewhere. In the Japanese tradition, for example, a single son inherits most property and becomes head of the family, with authority over other branches and multiple generations of couples living in the same household. So where the English male could leave the household and be free, Japan prioritized responsibility—a web of obligations of authority and sacrifice. Where England provided a fertile soil for liberalism, Japan was more fertile soil for authoritarian and conservative ideas.

23. This perhaps follows from Georg Hegel's assertion that reason was inseparable from the civilization in which it was found, rather than lying outside it, becoming natural, effortless, and part of what we are. For example, Pierre Bourdieu's notion of the *habitus* explained "the way society becomes deposited in persons in the form of lasting dispositions, or trained capacities and structured propensities to think, feel and act in determinant ways, which then guide them." Loïc Wacquant, "Habitus," in *International Encyclopedia of Economic Sociology*, ed. Jens Becket and Milan Zafirovski (London: Routledge, 2005), 316.

24. Stefana Broadbent and Mattia Gallotti, "Collective Intelligence: How Does It Emerge?" Nesta, accessed April 23, 2017, http://www.nesta.org.uk/sites/default/files/collective _intelligence.pdf.

25. Francis Galton described the crowd at a county fair accurately guessing the weight of an ox when their individual guesses were averaged. This figure was closer to the true weight than the estimates of individual members of the crowd, including the experts. Galton's anecdote has been repeated in many books, including James Surowiecki *The Wisdom of Crowds*.

26. To be fair, its main advocate, James Q. Wilson, always admitted that the "broken windows" theory was an interesting speculation rather than based on evidence.

27. Alfred Chandler, *Strategy and Structure: Chapters in the History of the American Industrial Enterprise* (Cambridge, MA: MIT Press, 1962).

28. Lev Vygotsky, *Mind in Society: The Development of Higher Psychological Processes* (Cambridge, MA: Harvard University Press, 1980).

Chapter 10: Self-Suspicion and Fighting the Enemies of Collective Intelligence

1. A. Pickering, *Science as Practice and Culture* (Chicago: University of Chicago Press, 1992), 54.

2. See Robert M. Galford, Bob Frisch, and Cary Greene, *Simple Sabotage: A Modern Field Manual for Detecting and Rooting Out Everyday Behaviors That Undermine Your Workplace* (New York: HarperOne, 2015).

3. For a demonstration of this in relation to politics, see Charles S. Taber and Milton Lodge, "Motivated Skepticism in the Evaluation of Political Beliefs," *American Journal of Political Science* 50, no. 3 (2006): 755–69.

4. Link to the *Guardian* article, featuring a video record of Spicer using the phrase: https://www.theguardian.com/us-news/2017/jan/23/sean-spicer-white-house-press-briefing-inauguration-alternative-facts.

5. Royal Institution Lecture on Mental Education (May 6, 1854), as reprinted in Michael Faraday, *Experimental Researches in Chemistry and Physics*, 1859, 474–75.

6. Pierre Bourdieu, *Distinction: A Social Critique of the Judgement of Taste* (London: Routledge, 1984), 471. This doxa may of course sometimes help us; Albert Hirschman wrote of the "hiding hand" that shields difficulties from us, which if we saw clearly, would stop us from setting out on challenging tasks.

7. John A. Meacham, "Wisdom and the Context of Knowledge: Knowing That One Doesn't Know," in *On the Development of Developmental Psychology*, ed. Deanna Kuhn and John A. Meacham (Basel, Swit.: Karger Publishers, 1983), 120.

8. It has to be chilly to work. Like a child, an institution can all too easily learn the wrong lessons or habits, or can learn better how to do bad things. Political science writes of the "we have learned" mantra, used after crises to restore legitimacy, but often without much real sign that lessons have been learned.

9. I'm not sure how a computer could acquire a habit of self-doubt; it could and frequently does self-verify, checking its facts against other facts. But in our traditions, doubt is deeper than this, serving as an often-recursive strategy of asking questions and remaining unsatisfied with the answers.

10. Though cognitive science continues to see human thought in terms of representations.

11. Bruno Latour, "Tarde's Idea of Quantification," in M. Candea, *The Social after Gabriel Tarde: Debates and Assessments* (New York: NY, Routledge, 2015).

12. For an interesting overview of the role of power in shaping data, see Miriam Posner, "The Radical Potential of the Digital Humanities: The Most Challenging Computing Problem

Is the Interrogation of Power," LSE Impact Blog, accessed April 24, 2017, http://blogs.lse.ac.uk /impactofsocialsciences/2015/08/12/the-radical-unrealized-potential-of-digital-humanities/.

13. For details on these examples, see Naomi Oreskes and Erik M. Conway, *Merchants of Doubt: How a Handful of Scientists Obscured the Truth on Issues from Tobacco Smoke to Global Warming* (London: Bloomsbury, 2012).

14. Overstanding is Raymond Tallis's phrase for the tendency of business-related and other books to overhype, exaggerate, and overclaim.

15. Igor Santos, Igor Miñambres-Marcos, Carlos Laorden, Patxi Galán-García, Aitor Santamaría-Ibirika, and Pablo García Bringas, "Twitter Content-Based Spam Filtering," 2014, accessed April 24, 2017, https://pdfs.semanticscholar.org/a333/2fa8bfbe9104663 e35f1ec41258395238848.pdf.

16. David Auerbach, "It's Easy to Slip Toxic Language Past Alphabet's Toxic-Comment Detector," MIT Technology Review, February 24, 2017, accessed April 24, 2017, https:// www.technologyreview.com/s/603735/its-easy-to-slip-toxic-language-past-alphabets-toxic -comment-detector/.

Chapter 11: Mind-Enhancing Meetings and Environments

1. For a useful recent analysis of how groups make good decisions, see Cass Sunstein and Reid Hastie, *Wiser: Getting beyond Group Think to Make Groups Smarter* (Cambridge, MA: Harvard Business Review Press, 2015). This book reaffirms the finding that it's often sensible to get groups to agree on a diagnosis before moving on to prescription.

2. Ones that meet these criteria include Future Search, 21st Century Town Meetings, and Design Thinking. See Steven M. Ney and Marco Verweij, "Messy Institutions for Wicked Problems: How to Generate Clumsy Solutions," accessed April 24, 2017, http:// www.stevenney.org/resources/Publications/SSRNid2382191.pdf. For a summary of the principles behind Future Search by its developers, see Martin Weisbord and Sandra Janoff, *Don't Just Do Something, Stand There! Ten Principles for Leading Meetings That Matter* (Oakland, CA: Berrett-Koehler Publishers, 2007).

3. "Estimate the Cost of a Meeting with This Calculator," *Harvard Business Review*, January 11, 2016, accessed April 24, 2017, https://hbr.org/2016/01/estimate-the-cost-of-a -meeting-with-this-calculator.

4. Harold Garfinkel, *Studies in Ethnomethodology* (Cambridge, UK: Polity Press, 1984); Erving Goffman, *The Presentation of Self in Everyday Life* (New York: Anchor Books, 1959); Michael Mankins, Chris Brahm, and Gregory Caimi, "Your Scarcest Resource," *Harvard Business Review* 92, no. 5 (2014): 74–80.

5. Ali Mahmoodi, Dan Bang, Karsten Olsen, Yuanyuan Aimee Zhao, Zhenhao Shi, Kristina Broberg, Shervin Safavi, Shihui Han, Majid Nili Ahmadabadi, Chris D. Frith, Andreas Roepstorff, Geraint Rees, and Bahador Bahrami, "Equality Bias Impairs Collective Decision-Making across Cultures," *Proceedings of the National Academy of Sciences of the United States of America* 112, no. 12 (2015): 3835–40.

6. There are many apps now available that support preparation and communication before meetings. Some, such as Do, include features that collaboratively build the agenda beforehand and send out automatic meeting notes. Others are specific to one challenge such as scheduling. Pick will find mutual availability between participants and then automatically book a convenient time for a meeting.

7. Hugo Mercier and Dan Sperber, "Why Do Humans Reason? Arguments for an Argumentative Theory," *Behavioral and Brain Sciences* 34, no. 2 (2011): 57–74.

8. Ray Dalio, *Principles*, accessed April 25, 2017, https://www.principles.com/#Principles.

9. Parmenides Eidos is a software program that visualizes complex data in more succinct ways to aid better decision making. See "Private and Public Services," Parmenides Eidos, accessed April 25, 2017, https://www.parmenides-foundation.org/application/parmenides-eidos/.

10. Anita Williams Woolley, Christopher F. Chabris, Alex Pentland, Nada Hashmi, and Thomas W. Malone, "Evidence for a Collective Intelligence Factor in the Performance of Human Groups," *Science* 330, no. 6004 (2010): 686–88.

11. Desmond J. Leach, Steven G. Rogelberg, Peter B. Warr, and Jennifer L. Burnfield, "Perceived Meeting Effectiveness: The Role of Design Characteristics," *Journal of Business and Psychology* 24, no. 1 (2009): 65–76.

12. Edward de Bono, *Six Thinking Hats* (Boston: Little, Brown and Company, 1985).

13. See "The Structural Dynamics Theory," Kantor Institute, accessed April 25, 2017, http://www.kantorinstitute.com/fullwidth.html.

14. Roughly speaking, the meeting mathematics formula runs as follows: meeting quality = [time × common grounding × relevant knowledge and experience] / [numbers × topic breadth].

15. For an instance of social network analysis applied to roles in meetings, see Nils Christian Sauer and Simone Kauffeld, "The Ties of Meeting Leaders: A Social Network Analysis," *Psychology* 6, no. 4 (2015): 415–34.

16. Joseph A. Allen, Nale Lehmann-Willenbrock, and Nicole Landowski, "Linking Pre-Meeting Communication to Meeting Effectiveness," *Journal of Managerial Psychology* 29, no. 8 (2014): 1064–81.

17. Steven G. Rogelberg, Joseph A. Allen, Linda Shanock, Cliff Scott, and Marissa Shuffler, "Employee Satisfaction with Meetings: A Contemporary Facet of Job Satisfaction," *Human Resource Management* 49, no. 2 (2010): 149–72.

18. Like Nesta's "randomized coffee trial," which encourages people to meet people they don't know in the workplace and has now been adopted by many big employers.

19. In everyday operational meetings, as much as strategic and creative events, meetings provide a place for participants to demonstrate their vision and mission. Joseph A. Allen, Nale Lehmann-Willenbrock, and Steven G. Rogelberg, *The Cambridge Handbook of Meeting Science* (Cambridge: Cambridge University Press, 2015).

20. For one approach that emphasizes this last capacity to synthesize from complexity, see Stafford Beer, *Beyond Dispute: The Invention of Team Syntegrity* (Chichester, UK: John Wiley, 1994); Markus Schwaninger, "A Cybernetic Model to Enhance Organizational Intelligence," *Systems Analysis, Modeling, and Simulation* 43, no. 1 (2003): 53–65.

21. The work of the Affective Computing Group at the Massachusetts Institute of Technology has been promising on this, with its focus on how digital technologies can better communicate emotions, although it raises as many questions as it answers, such as when hostility is or isn't good for meetings, or whether mutual transparency improves the quality of decisions or instead fuels conformism.

22. Amy MacMillan Bankson, "Could an Artificial Intelligence-Based Coach Help Managers Master Difficult Conversations?" MIT Sloan School of Management, February

23, 2017, accessed April 26, 2017, http://mitsloan.mit.edu/newsroom/articles/could-an-ai
-based-coach-help-managers-master-difficult-conversations/.

23. William T. Dickens and James R. Flynn, "Heritability Estimates versus Large Environmental Effects: The IQ Paradox Resolved," *Psychological Review* 108, no. 2 (2001): 346–69.

24. Source: Michael Weiser, "The Computer for the 21st Century," accessed at: https://www.ics.uci.edu/~corps/phaseii/Weiser-Computer21stCentury-SciAm.pdf.

25. Christian Catalini, "How Does Co-Location Affect the Rate and Direction of Innovative Activity?" *Academy of Management Annual Meeting Proceedings* 1 (2012): 1.

Chapter 12: Problem Solving: How Cities and Governments Think

1. The London Collaborative was led by the Young Foundation (where I was chief executive), and also involved the Office of Public Management and Common Purpose.

2. This was Boris Johnson. To be fair, he had little experience with any kind of management and so was unlikely to understand the need for tools of this kind.

3. The city needs to organize three types of knowledge: what's proven (actions for which there is as solid evidence base), what's promising (and therefore warrants testing and development), and what's possible (the more imaginative options that require hard thought and design to bring them to birth). See Nesta's work setting up the Alliance for Useful Evidence, the creation of new What Works centers, and detailed work on what it means for something to be proven.

4. For now, smart city projects usually involve simplifying the messy complexity of the city into something that approximates an engineering diagram. An opposite strategy deliberately cultivates complexity—differentiated neighborhoods, hybrid transport systems, food economies, and so on, and is likely to be a better route to genuine collective intelligence, given the complexity of the tasks that cities have to deal with.

5. Constraints also help with creativity—and some of the best-known tools for problem solving, such as TRIZ (developed in the USSR in the 1940s and then adopted by engineers globally) deliberately make the most of constraints to accelerate thinking.

6. The fast idea generator (www.diy.org) that I developed offered a simple and universal language for creating ideas—a comprehensive set of processes with which any group can quickly multiply options.

7. Allan Afuah and Christopher L. Tucci, "Crowdsourcing as a Solution to Distant Search," *Academy of Management Review* 37, no. 3 (2012): 355–75.

8. Roger J. Hollingsworth, "High Cognitive Complexity and the Making of Major Scientific Discoveries," in *Knowledge, Communication, and Creativity*, ed. Arnaud Sales and Marcel Fournier (London: Sage, 2007), 134.

9. James G. March, "Exploration and Exploitation in Organizational Learning," *Organization Science* 2, no. 1 (1991): 86.

10. That's easier said than done. To illuminate the point, a classic paper in political science looked at the problem of school students failing to return their cafeteria trays. There were many competing theoretical explanations (well over thirty), but few obvious ways to judge which ones to use. Lloyd S. Etheredge, "The Case of the Unreturned Cafeteria Trays" (paper, American Political Science Association, Washington, DC, 1976), accessed April 26, 2017, http://www.policyscience.net/ws/case.pdf.

11. George Polya, *How to Solve It*, (Garden City, NY: Doubleday, 1957), 115.

12. Carl Sagan, *Cosmos* (New York: Ballantine Books, 1980), 218.

13. Though it will be obvious why artificial intelligence struggles with most complex problems, which require so much iteration between questions and answers, and so much zooming in and out.

14. Judea Pearl, "Causal Inference in Statistics: An Overview," *Statistics Surveys* 3 (2009): 96–146. For a discussion more relevant to public policy, see Tristan Zajonc, "Essays on Causal Inference for Public Policy" (PhD diss., Harvard University, 2012).

15. In Thomas Hobbes's account, the state was akin to an automaton, a clockwork machine, a combination of mind and body.

16. Charles Sabel, *Learning by Monitoring* (Cambridge, MA: Harvard University Press, 2006).

17. Beth Noveck has been uniquely effective as both a pioneer and observer. See Beth Noveck, *Smart Citizens, Smarter State: The Technologies of Expertise and the Future of Governing* (Cambridge, MA: Harvard University Press, 2015).

18. This report from a few years ago showed how social network analysis could map the patterns of partnership in a town or city, revealing the human reality of cooperation. Nicola Bacon, Nusrat Faizullah, Geoff Mulgan, and Saffron Woodcraft, "Transformers: How Local Areas Innovate to Address Changing Social Needs," Nesta, January 2008, accessed April 28, 2017, http://www.maximsurin.info/wp-content/uploads/pdf/transformers.pdf. These tools have yet to become widely used, but are relatively cheap and easy to implement.

19. See Geoff Mulgan, "Innovation in the Public Sector: How Can Public Organisations Better Create, Improve, and Adapt?" Nesta, November 2014, accessed April 28, 2017, http://www.nesta.org.uk/sites/default/files/innovation_in_the_public_sector-_how_can_public_organisations_better_create_improve_and_adapt_0.pdf; Ruth Puttick, Peter Baeck, and Philip Colligan, "I-Teams: The Teams and Funds Making Innovation Happen in Governments around the World," June 30, 2014, http://www.nesta.org.uk/publications/i-teams-teams-and-funds-making-innovation-happen-governments-around-world.

10. Geoff Mulgan, *The Art of Public Strategy: Mobilizing Power and Knowledge for the Common Good* (Oxford: Oxford University Press, 2009).

21. I helped write a report that documents in more detail what these tools are and how they are being used around the world. This formed part of a project with the United Nations designed to upgrade how country strategies were contributing to achieving Sustainable development goals. Geoff Mulgan and Tom Saunders, "Governing with Collective Intelligence," Nesta, January 2017, accessed April 28, 2017, http://www.nesta.org.uk/sites/default/files/governing_with_collective_intelligence.pdf.

Chapter 13: Visible and Invisible Hands:
Economies and Firms as Collective Intelligence

1. More information can also bring in new biases and discrimination as well as illumination, like the evidence in New York that white Airbnb providers charge 12 percent more than black providers for equivalent accommodation. Benjamin Edelman and Michael Luca, "Digital Discrimination: The Case of Airbnb.com," HBS working paper series, January 10, 2014, accessed April 28, 2017, http://www.hbs.edu/faculty/Publication%20Files/Airbnb_92dd6086-6e46-4eaf-9cea-60fe5ba3c596.pdf.

2. F. Hayek, *Individualism and Economic Order* (Chicago, IL: University.of Chicago Press, 1948), 87.

3. Li and Fung is one of the biggest companies few have heard of. It links manufacturers and retailers, organizing logistics, taxes, supply chains, and design, and has evolved into something quite like the assemblies I've described earlier.

4. Baruch Lev and Feng Gu, *The End of Accounting and the Path Forward for Investors and Managers* (Chichester, UK: John Wiley, 2016).

5. This model, pioneered by Nesta, has now been adopted by the UK government and is part of a broader program for measuring cognition in the economy. The work has been led by Hasan Bakhshi, and resulted in a series of publications on measuring the creative economy and, more recently, the regular publication of statistics by the Department for Culture, Media, and Sport in the United Kingdom.

6. From his Nobel Prize acceptance speech, collected in: Sture Allen, *Nobel Lectures in Literature, 1968–1980* (London: World Scientific Publishing Company, 2007), 135

7. One example of this is industrial policy, where governments are increasingly interested in how to cultivate new dynamic comparative advantages that are unlikely to be caught by economic indicators (which tend to reflect past structures of the economy). Instead, the key indicators are ones that capture patterns of emergence (such as firm creation, or which places are creating new clusters, new competences, and market demands). Consistency over time may be less useful than pattern recognition, since measures that captured well a previous economic era (for example, based on large-scale manufacturing) will be too blunt to say much about emerging industries. The same is true of many social issues, where the key indicators are ones that show up clusters and cultural patterns. These are not easily captured by overall poverty measures; rather, as in industrial policy, often the greatest insights come from spotting clusters and spatial concentrations. Again, although consistent time series data may be interesting, they risk missing the most important issues for policy makers. Greater attention to behavioral issues is pushing in the same direction, since meaningful observations about behavior are quite rare at too aggregated a level. In contrast, anyone wanting to understand behavior has to segment populations, such as with the many descendants of the Stanford Lifestyle measures (Arnold Mitchell developed the Values and Lifestyles psychographic methodology at Stanford to explain changing US values and lifestyles). New knowledge from genetics is further reinforcing this trend toward disaggregation along with a view of economies as more cellular, networked, and fractal in some respects, but not guided by the grand aggregates that so dominated the mental landscape of the second half of the twentieth century.

8. Another argument says that this is the job or choice of shareholders. They are where the buck stops. Certainly there are good reasons for wanting shareholders to become more intelligent users of their power, not just to stop foolish mergers and half-baked business strategies, but also to introduce greater scrutiny of social and environmental outcomes. Shareholders should be on the right side of this assertion—but they too can't be relied on to police evidence, any more than political parties in parliaments can be relied on to keep the governments they support honest about it.

Chapter 14: The University as Collective Intelligence

1. Quoted in Geoff Mulgan, Oscar Townsley, and Adam Price, "The Challenge-Driven University: How Real-Life Problems Can Fuel Learning," Nesta, accessed April 28, 2017, https://www.nesta.org.uk/sites/default/files/the_challenge-driven_university.pdf.

2. Derek Bok has made a similar argument in various books, drawing on his own experience running several top US universities.

3. Some of the mistakes can be attributed to the naïveté of the computer scientists and venture capitalists who made much of the running, and this may not matter, except to investors who will lose their money. The best-funded ones may fight their way through to significant impact or succeed as perfectly benign marketing tools for the most established universities.

4. Geoff Mulgan, Oscar Townsley, and Adam Price, "The Challenge-Driven University: How Real-Life Problems Can Fuel Learning," Nesta, accessed April 28, 2017, https://www .nesta.org.uk/sites/default/files/the_challenge-driven_university.pdf. A recent book coauthored by Clayton Christensen, *The Innovative University: Changing the DNA of Higher Education from the Inside Out* (Hoboken, NJ: John Wiley and Sons, 2011), unintentionally showed the depth of the problem. Christensen is an impressive writer and thinker whose most famous idea—disruptive innovation—is broadly sound. Yet it can't be a coincidence that this is his weakest book. It offers interesting accounts of the history of two US universities, but struggles to address basic questions about what universities are or should be for. It only discusses innovation in relation to teaching, and simply doesn't mention the great majority of recent innovations, except for a handful of examples from US universities. Crucially, it has literally nothing to say about how higher education could become more systematic in its innovation.

5. Funding is needed to support experiments that address how the key elements of the university could work better: knowledge generation, discipline formation, skills transmission, social network forming, and connected economic development, preferably in ways that can support outsiders as well as insiders with promising ideas. We might expect some of these to focus on such things as emerging disciplines, from computational social science to social epigenetics. Others might address new ways of thinking and learning—like "studio" methods of working around problem solving in teams, with real-life partners (as pioneered by Aalto); new methods of knowledge generation coming from the open science movement; new ways of turning knowledge into useful forms such as labs and accelerators; new ways of keeping costs down (like South Africa's CIDA Empowerment Fund); or new ways of rethinking the role of the university in relation to life stage, such as Harvard's Advanced Leadership Initiative or the University of the Third Age movement. A high priority in some countries would be innovation to help universities advance social mobility, since many higher education systems now do the opposite.

Chapter 15: Democratic Assembly

1. Thomas Hobbes, *Leviathan* (1651; repr., London: Penguin, 1982), I.16.13.

2. Christopher Achen and Larry Bartels, *Democracy for Realists* (Princeton, NJ: Princeton University Press, 2016).

3. Letter from John Adams to John Taylor, December 17, 1814, in *Founders Online, National Archive*, accessed at: https://founders.archives.gov/documents/Adams/99-02-02 -6371.

4. Martin Gilens and Benjamin I. Page, "Testing Theories of American Politics: Elites, Interest Groups, and Average Citizens," *Perspectives on Politics* 12, no. 3 (2014): 575.

5. Voting behavior is instead better understood as an aspect of social expression. It is as much micro as macro, about how we interact with neighbors, families, and friends, and the broad conclusion of research on attempts to drive voting levels up is that "the more personal the mode of contact, the more effective it is." Todd Rogers, Craig R. Fox, and Alan S. Gerber, "Rethinking Why People Vote: Voting as Dynamic Social Expression," in *The Behavioral Foundations of Policy*, ed. Eldar Shafir (Princeton, NJ: Princeton University Press, 2012), 91.

6. Notably in the work of Charles Lindblom, including in his seminal book *The Intelligence of Democracy* (New York: Free Press, 1965), and before him Dewey.

7. Fareed Zakaria's *The Future of Freedom* (New York: W. W. Norton, 2003) is a particularly interesting recent investigation of the subtle relationships between democracy and liberty, and echoes some of these earlier warnings in its advocacy of liberties ahead of full democracy.

8. As with any tool for collective intelligence, the detailed designs had to balance conflicting priorities. For example, the framers of the French Constitution after the revolution of 1789 rejected the English jury system's requirement for unanimity. They realized that a balance had to be struck between the risks of wrongful conviction and wrongful acquittal, but disagreed as to where the balance should lie. Pierre-Simon Laplace believed that a 30 percent chance of executing an innocent person was not acceptable; Siméon Denis Poisson argued that executing two innocent out of seven accused was reasonable and so recommended that juries should make decisions by simple majority. In the 1960s, the US Supreme Court declared it constitutional for juries to make a decision by simple majority, and that same year England allowed conviction by a ten-to-two majority. Similar judgments have to be made by any group or committee: when is consensus essential and when is a simple majority enough?

9. Source: http://press.princeton.edu/titles/10671.html.

10. Though none have adopted the proposal of Harvard's Otto Eckstein that when considering policy options, parliaments should be able to see how different weightings for such things as environmental benefits, jobs, or mobility would affect cost-benefit analysis, a transparent but democratic decision tool.

11. These experiments have been helped by new theoretical explorations, such as simulated models that show how agents come to decisions or consensus. See, for example, Thomas Seeley, *Honeybee Democracy* (Princeton, NJ: Princeton University Press, 2010).

12. Liz Barry, "vTaiwan: Public Participation Methods on the Cyberpunk Frontier of Democracy," Civicist, accessed April 29, 2017, http://civichall.org/civicist/vtaiwan -democracy-frontier/.

13. Recent experience suggests that overreliance on digital tools rather than print, radio, television, and face-to-face interaction can get skewed inputs. Even technologically literate cities like New York and Los Angeles have repeatedly found that participants in purely digital consultations are much more male, young, well educated, affluent, and metropolitan than the population as a whole. That may be acceptable for some kinds of engagement—1 percent involvement *can* greatly improve the quality of decisions. But the more successful methods combine online and off-line, digital and face-to-face.

14. Various experiments—say, enabling expatriates to take part in Colombia's referendum—point the way to a possible future where far-larger numbers influence global decisions. Democracy Earth, "A Digital Referendum for Colombia's Diaspora,"

accessed April 29, 2017, https://medium.com/@DemocracyEarth/a-digital-referendum
-for-colombias-diaspora-aeef071ec014#.x2km4cs93.

Chapter 16: How Does a Society Think and Create as a System?

1. Roberto Mangabeira Unger, *The Self Awakened: Pragmatism Unbound* (Cambridge, MA: Harvard University Press, 2007).

2. Some of this framework draws on recent work by Nesta: our work to devise new ways of mapping the world, including new tools combining open data, official statistics, and web-scraping to describe phenomena, such as the growth of new industries; our work on engaging people in the design of their own systems, such as through programs of "people-powered health" that engage patients in the codesign, cocreation, and comanagement of budgets and services; and our work on new structures within governments that aim to innovate systemically, including i-teams and labs.

3. These systems have some autopoietic qualities, but only to a limited degree. If we measure autopoiesis by how much of the complexity of a system is defined by itself and how much is defined by its environment, these are systems substantially defined by their environments: by rules, tariffs, regulations, and knowledge that exist at a national and even global level, and are imposed on local systems. For the system to adapt intelligently, we need to increase both its "partness" and "apartness"—its ability to draw on knowledge from a global commons (for instance, on what works in health treatments), and its sense of responsibility and apartness.

4. Community organizing often starts with a group of individuals living in particular places and conditions that do not yet have a shared identity. The organizers use shared and salient problems as the basis for creating a common identity, and the effort to solve these problems is useful instrumentally, but also as a way of creating a group identity and capacity that can then be used instrumentally to take on the next big issues.

5. This was one of the insights of Niklas Luhmann's work on subsystems and the incommensurability of their language.

6. In this case, we can focus on combining pathway innovation with innovation around new linking elements. These include common assessment tools and language, common triage methods, common protocols on data sharing, and common call centers and case tracking. Some external help may be crucial in designing these in collaboration with insiders. There are also new inputs (for example, supervolunteers) and new coordination mechanisms.

7. Action then moves on to options for change. Some are about alignment—and the use of the new linking tools described above, which help the parts of the system to connect more effectively; some aim at minimum necessary alignment rather than complete alignment; some are about generating new deals and microcollaborations—with multiple interactions to generate these; and some involve structural recursion with microexamples of the bigger system (such as the individual case, surrounded by its microsystem of supports).

8. Once again in this example, new commons emerge combining data, information, knowledge, and judgment. A key insight is that these are likely to be underprovided—with a lack of institutions with the resources, incentives, or skills to fulfill these roles (these types of common are also reflexive, with links between micropools and macro, the microcommons being such things as the study circle or the conversation between patient and doctor that mirrors the formal knowledge system).

9. This is the R-UCLA scale, which asks about a participant's feelings related to loneliness, such as "How often do you lack companionship?," and assigns an overall loneliness score (from four to twelve).

10. Though we can also layer onto these more subtle feedback tools to help the system know itself: tracking sentiment, anxiety, or commitment in real time for the systems community, or using social network analysis–type tools to study who is helpful to whom across networks of collaboration.

Chapter 17: The Rise of Knowledge Commons: It's for Everyone

1. In science, which has thrived for two centuries largely as a commons, there are the movements to open up raw research data and all research findings to public availability within a year of publication. The movement to open up the data generated by publicly provided or funded activities links many of these themes together, and has turned what used to be an internal resource for governments and businesses into a commons, prompting rapid progress in some fields, such as transport. Many other examples have built on enthusiasm and value, like the spread of creative commons and other open approaches to intellectual property, or offering tools like WordPress as open-source resources.

2. The idea that massaging Wagyu cows improves their quality turns out to be a myth—but it's a nice one.

3. The Internet was created in the United States and many of its pioneers had a strong commitment to providing it as a commons. But despite exceptions like the open-source movement, they struggled to find ways of translating that spirit into viable economic forms. If the Internet had emerged in Europe, things might be different (though not necessarily better). After all, Europe pioneered public service broadcasting, with its various missions to educate and inform as well as entertain, just as in a previous period it pioneered the idea of free museums or science provided freely for the common good. In the age of the Internet, though, Europe has pioneered nothing comparable, with the partial exception of Skype. And of course in other parts of the world, this is an age of tightening controls rather than freedom, where the new commons are threatened by power, whether that is state suppression (Russia and China), corporate power over information (India), or fear from organized crime (Mexico).

4. One attempt to provide answers was Destination Local (http://www.nesta.org.uk /project/destination-local), a Nesta program. The Knight Foundation has acted in a similar way, although on a much larger scale, in the United States.

5. See John Loder, Laura Bunt, and Jeremy C. Wyatt, "Doctor Know: A Knowledge Commons in Health," Nesta, March 11, 2013, accessed May 2, 2017, http://www.nesta .org.uk/publications/doctor-know-knowledge-commons-health.

6. The recent development of OAuth as an open standard is a good illustration of the emergence of a new commons, encouraging take-up because it is both free and more reliable, having been scrutinized by more experts than proprietary equivalents. Google, LinkedIn, and Twitter, among others, now use it as a de facto global standard. Other promising examples include models like Open Mustard Seed at the Massachusetts Institute of Technology and Mydex in the United Kingdom.

7. A recent work in progress is the attempt to provide an independent source of guidance for teachers on what technologies to buy; there are powerful incentives for marketing

technologies, but far weaker incentives for anyone to appraise how well they work. The former is a private good, and the latter is a commons.

8. "Engaging News about Congress: Report from a News Engagement Workshop," Engaging News Project, accessed May 2, 2017, https://engagingnewsproject.org/.

9. For a good and impassioned dissection of this in relation to the 2016 US presidential election, see Joshua Benton, "The Forces That Drove This Election's Media Failure Are Likely to Get Worse," NiemanLab, November 9, 2016, accessed May 2, 2017, http://www.niemanlab.org/2016/11/the-forces-that-drove-this-elections-media-failure-are-likely-to-get-worse/.

10. At least four families of commons look set to emerge. One group will provide a safe haven for personal data, allowing people to choose who gets to see which items of data about themselves and their choices. A second group will be commons that combine public data—say, about the weather or economy—in ways that are most useful. A third group will combine data about motion and coordination—for example, the position of drones in a city. A fourth might combine knowledge about a field—such as health care—in ways that can be linked to personal information—for instance, from wearable devices or genetic tests. Each of these will need its own rules of governance and economic base. Some may emerge organically through crowdfunding, pledge banks, and other devices that combine free choice with collective action. But these are unlikely to be adequate.

Chapter 18: Collective Wisdom and Progress in Consciousness

1. Though other estimates are much more modest. See Melanie Arntz, Terry Gregory, and Ulrich Zierhahn, "The Risk of Automation for Jobs in OECD Countries: A Comparative Analysis," OECD Social, Employment, and Migration Working Papers 189, 2016, accessed May 2, 2017, http://www.ifuturo.org/sites/default/files/docs/automation.pdf.

2. Some thoughtful economists now argue that "this time is different," and that even if the warnings were wrong in the past, they're now right. The nature of communications and information technologies means that they really will scythe through professional jobs. They may be right. But given how wrong comparable analyses have been in the past, we should start from a position of skepticism. Societies have often in the past created novel roles to satisfy people's need for recognition, whether to keep aristocrats busy in medieval Europe or the unemployed busy in the 1930s' United States. It's possible that we will see comparable moves.

3. For this argument, see Stephen Hsu, "Don't Worry, Smart Machines Will Take Us with Them: Why Human Intelligence and AI Will Co-evolve," *Nautilus*, September 3, 2015, accessed May 2, 2017, http://nautil.us/issue/28/2050/dont-worry-smart-machines-will-take-us-with-them.

4. From *Civil Elegies* (Toronto: House of Anansi Press, 1972), 56. This was paraphrased for the walls of the newly built Scottish Parliament.

5. J. Maynard Smith and E. Szathmáry, E. *The Major Transitions in Evolution* (Freeman, Oxford, 1995).

6. Yuval Noah Harari's book *Homo Deus: A Brief History of Tomorrow* (New York: Harper, 2015) warns of the rise of "dataism" as a new religion that stands in the shoes of data versus the universe. Much of his description is accurate, but it is of a thin belief system unlikely to satisfy many.

7. Spiral dynamics was a theory originally devised by Graves (1914–86), professor emeritus in psychology at Union College in New York. The ideas were then taken further by Don Beck and Chris Cowan in *Spiral Dynamics: Mastering Values, Leadership, and Change* (Hoboken, NJ: Blackwell Publishers, 1996), and linked to the integral theories of Ken Wilber in *A Theory of Everything: An Integral Vision for Business, Politics, Science, and Spirituality* (Boulder, CO: Shambhala Publications, 2000). Around them have clustered a group of management consultants, academics, and professionals in organizational development. These theories have won many adherents, but have some weaknesses, including the lack of verification, internal contradictions, lack of attention to what's known about the topics being discussed, and the careless use of terms. For a useful and tough review of a recent book in this tradition, see Zaid Hassan, "Is Teal the New Black? Probably Not," Social Labs Revolution, July 13, 2015, accessed on May 2, 2017, http://www.social-labs.com/is-teal-the-new-black/.

8. Also attributed to Eden Phillpotts, *A Shadow Passes* (London: Cecil Palmer & Hayward, 1918), 19.

Afterword: The Past and Future of Collective Intelligence as a Discipline

1. In Gary Wills, *Why Priests* (New York, NY: Viking, 2013), 120.

2. Pierre Levy, *Collective Intelligence: Mankind's Emerging World in Cyberspace* (New York: Basic Books, 1999).

3. Howard Bloom, *Global Brain: The Evolution of Mass Mind from the Big Bang to the 21st Century* (New York: John Wiley and Sons, 2001).

4. Jie Ren, Jeffrey V. Nickerson, Winter Mason, Yasuaki Sakamoto, and Bruno Graber, "Increasing the Crowd's Capacity to Create: How Alternative Generation Affects the Diversity, Relevance, and Effectiveness of Generated Ads," *Decision Support Systems* 65 (2014): 28–39.

5. Enrico Coen, *Cells to Civilizations: Principles of Change That Shape Life* (Princeton, NJ: Princeton University Press, 2012).

6. Hélène Landemore and Jon Elster, eds., *Collective Wisdom: Principles and Mechanisms* (New York: Cambridge University Press, 2012).

7. Mancur Olson, *The Logic of Collective Action: Public Goods and the Theory of Groups* (Cambridge, MA: Harvard University Press, 1971).

8. In Mary Douglas, *How Institutions Think* (Syracuse, NY: Syracuse University Press, 1986).

9. Anita Williams Woolley, Christopher F. Chabris, Alex Pentland, Nada Hashmi, and Thomas W. Malone, "Evidence for a Collective Intelligence Factor in the Performance of Human Groups," *Science* 330, no. 6004 (2010): 686–88; Dirk Helbing, "Managing Complexity," in *Social Self-Organization: Understanding Complex Systems*, ed. Dirk Helbing (Berlin: Springer, 2012), 285–99.

10. For another useful source, see Thomas W. Malone and Michael S. Bernstein, *Handbook of Collective Intelligence* (Cambridge, MA: MIT Press, 2015).

11. Some examples include Nesta's wide-ranging work on data, work in health such as Dementia Citizens, D-CENT in democracy, and collective intelligence in development.

INDEX